POWER IN NUMBERS

Quarto Knows

Inspiring | Educating | Creating | Entertaining

Brimming with creative inspiration, how-to projects, and useful information to enrich your everyday life, Quarto Knows is a favorite destination for those pursuing their interests and passions. Visit our site and dig deeper with our books into your area of interest: Quarto Creates, Quarto Cooks, Quarto Homes, Quarto Lives, Quarto Drives, Quarto Explores, Quarto Gifts, or Quarto Kids.

First published in 2018 by Race Point Publishing, an imprint of The Quarto Group, 142 West 36th Street, 4th Floor, New York, NY 10018, USA
T (212) 779-4972 **F** (212) 779-6058 **www.QuartoKnows.com**

Race Point titles are also available at discount for retail, wholesale, promotional, and bulk purchase. For details, contact the Special Sales Manager by email at specialsales@quarto.com or by mail at The Quarto Group, Attn: Special Sales Manager, 401 Second Avenue North, Suite 310, Minneapolis, MN 55401, USA.

10 9 8 7 6 5 4 3 2 1

ISBN: 978-1-63106-485-2

Editorial Director: Jeannine Dillon
Acquiring Editor: Melanie Madden
Managing Editor: Erin Canning
Cover Illustrations for Eugenia Cheng,
Emmy Noether, and Annie Easley: Sean Yates
Cover Design: Merideth Harte
Interior Design: Jen Cogliantry

Printed in China

POWER IN NUMBERS

THE REBEL WOMEN OF MATHEMATICS

TALITHIA WILLIAMS, PhD

Race Point
PUBLISHING

CONTENTS

INTRODUCTION

Above: Williams on the campus of Harvey Mudd College.

Sometimes it's hard for people to hide their shock when I tell them I'm a mathematician. I get it. After all, I didn't know any women with PhDs in mathematics growing up in Columbus, Georgia. Those same people might ask me how I got here. I suppose for me it started in high school. I qualified for five Advanced Placement (AP) classes and, while I would like to say that the reasons I signed up for those AP courses were honorable ("I wanted a rigorous academic environment" or "I wanted to be challenged with college-level material"), the reality was, as soon as the teacher told me that making a "B" in AP was like getting an "A" in the standard curriculum, I was sold.

It was when I was taking AP Calculus that marked the first moment I considered making math my life. In a high school with well over two thousand students, there was only one AP Calculus class, with about twenty-five kids. Of those twenty-five students, only four of us were African American. My teacher, Mr. Dorman, was in his early fifties at the time. He would invite us up to the chalkboard to work out problems for extra credit. I always went up there because I wanted that extra credit. I wasn't a star student, but I was certainly motivated for a little extra credit. Who wouldn't be?

One day after class, Mr. Dorman pulled me aside and told me he thought I was talented in math, and that I should think about majoring in it when I went to college. As a seventeen-year-old kid, I was shocked to hear him say that. I mean, sure, my mom and dad told me I was smart, but they're my parents; I have their genes. But Mr. Dorman was different. He was a teacher, and his encouragement would forever change my life trajectory. Actually, it was the first time an older white man affirmed my intellectual ability. Even though I never saw myself as a mathematician, he saw me as one. The conversation changed me. It changed my life.

So, you know what I did? I went to college . . . and I majored in math. The funniest part of the story is that at my ten-year high-school reunion, I discovered that Mr. Dorman had said the same thing to many of his students, but by then, the damage had

Below: Williams presenting her popular TED talk, "Own Your Body's Data," at the 2017 Tableau Conference.

already been done! And I don't regret my choice for a second. It just goes to show you that the power of a teacher who believes in you can be transformational. They can act as a mirror for potential, even if you can't yet see it in yourself.

As an undergraduate student, I spent three summers working at NASA's Jet Propulsion Laboratory (JPL). At JPL I was assigned to a research team where Dr. Lonne Lane was the advisor. Lonne was ridiculously brilliant, and he insisted that I call him by his first name. Now being raised in the South, I was taught never to call adults by their first name. Some social title needed to come before his name—Mr., Dr., Officer, Sir, Rev, Deacon—as a sign of respect. But he wouldn't let me do it. In fact, everyone at JPL was on a first-name basis. He brought me into his research team and made me feel like I was an asset, as if my thoughts and opinions mattered. Here I was, this kid from Georgia, working side by side with rocket scientists. But despite the camaraderie, I still couldn't help but feel a bit out of place. That is, until I met Dr. Claudia Alexander.

Claudia was my mentor for those three summers. She was everything that I dreamed I could become. She was beautiful and smart, and had a PhD in space physics from the University of Michigan, and worked at NASA. I remember wanting to get beautiful blonde highlights just like hers. The better I got to know her, the more I became hooked on mathematics. You see, she represented the dream of what I could one day become. I didn't know it then, but during our summers together at NASA, she had planted a seed in me, and it was, thankfully, a lot more rewarding than blonde highlights!

I graduated from Spelman College as a math major with a minor in physics, before moving on to Howard University for a master's degree in math. At Rice University, where I received my PhD in statistics, I was not only the only female in my class, but also the only African American. I met my husband while at Rice—he was in the Computational and Applied Mathematics department—and he approached me with a line like no other: "I wish I was your derivative, because then I would be tangent to your curves." Yeah, who wouldn't pass that up? He had me at "derivative."

A few years later, I landed a tenure-track position at Harvey Mudd College and have worked to make the dream of broadening participation in the mathematical sciences a reality, much like Mr. Dorman, Claudia Alexander, Lonne Lane, and my Spelman professors have done for me. And I hope *Power in Numbers* takes me one step closer to that dream.

As you read through the incredible stories of the women profiled in these pages, the one thread they all have in common is that at some point on their journeys, someone believed in them; someone made them think the impossible was perhaps not so impossible. It was extremely difficult to narrow down the list of women I selected for inclusion here, but what I offer is a sampling of female mathematicians throughout history—including modern mathematicians—who have shattered stereotypes, pursued their passions, and persisted even when things got tough—even when people told them they couldn't do it. Many were alone on their journey, but with every female who enters the field of math, it makes it easier and more achievable for the next one, and the one after that.

I invite women and men of all nationalities and backgrounds to learn about these dynamic mathematicians and scientists that have shaped our society. May their stories empower the next generation of STEM rebels to conitnue advancing mathematical theory, bringing awareness to the field, and increasing our Power in Numbers.

Above: Talithia Williams, co-host of NOVA WONDERS.
Courtesy / © 2018 WGBH Educational Foundation.

PART 1
The Pioneers

The field of mathematics is not known for being especially friendly or appealing to women throughout history, but with the explosion of sophisticated technology in the twentieth century and beyond, many female mathematicians are making essential contributions to all kinds of human endeavors, from bioinformatics to spaceflight. The rise of universities in Ancient Greece laid the groundwork for modern university education and eventually for the breakthroughs of today, with teachers presenting subjects like philosophy and rhetoric alongside mathematics and astronomy. One such professor was Theon of Alexandria (ca. 335–ca. 405), who taught mathematics at the local university and wrote commentaries on some of the greatest scientific works of Ancient Greece, including Euclid's *Elements* and Ptolemy's *Almagest*. Theon passed his passion for learning to his daughter Hypatia (d. 415), who helped him write those commentaries and whose reputation would surpass his own.

Opposite: Emmy Noether

Hypatia studied philosophy, astronomy, and mathematics in Athens, Greece, before she became head of the Neoplatonist school in Alexandria around the turn of the fifth century. She was a particularly skilled speaker, and many people from other cities flocked to the intellectual center of Egypt to hear her lectures. Within mathematics, she is best

Above: When a double cone is intersected by a plane, the parabola (left), ellipse (center), and hyperbola (right) are created.

known for her work on Apollonius's treatise on conic sections, which divided cones into different parts using flat geometric planes and introduced the concepts of the parabola, hyperbola, and ellipse.

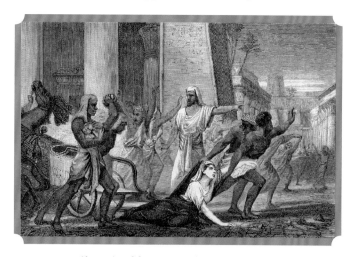

Above: An 1865 engraving depicting the death of Hypatia.

Like too many women in this book, Hypatia paid a heavy price for her pursuit of knowledge, though her story is the most gruesome. As recounted in the 2009 film *Agora* (and countless plays and works of fiction throughout history), the Bishop of Alexandria spread virulent rumors about her, and one day in 415, she was attacked by a Christian mob, stripped, stabbed to death with broken pottery, and dragged through the streets.

It is almost certain that other female mathematicians quietly carried on her legacy through the European Medieval period, the Chinese Song Dynasty, the Islamic Golden Age, and the Renaissance, but there is little record of female achievements in mathematics prior to the publication of French mathematician and philosopher René Descartes's monumental treatise *Discourse on the Method*—original source of the famous expression cogito ergo sum ("I think, therefore I am")—in 1637. Just four years later, a French woman of obscure origins named Marie Crous published a study on the decimal numerical system, which introduced the decimal point that is central to our modern notation. A hundred years after that, Frenchwoman Émilie du Châtelet (1706–1749) published *Institutions de Physique* (*Foundations of Physics*), which explained and analyzed the cutting-edge mathematical ideas of

Above: Detail of an eighteenth-century painting of Du Châtelet at her desk.

In mathematics, Agnesi was best known for her illustration of the "Witch of Agnesi," a type of versed sine curve originally studied by French mathematician Pierre de Fermat. Her textbook used the term *versiera* (from the Latin *vertere*—"to turn"), but when it was translated into English, *versiera* was taken to be an abbreviation for *avversiera*, the Italian word for "wife of the devil." The name stuck.

The curve is commonly expressed by the Cartesian equation $y = 8a^3/(x^2 + 4a^2)$, where a is equal to the radius of the circle in the following illustration.

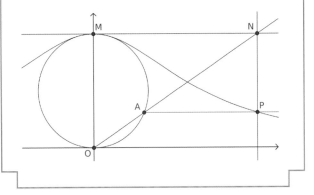

Gottfried Leibniz, who later became co-discoverer of calculus, with Sir Isaac Newton. Naturally rumors spread that Châtelet was merely rehashing the lessons of her tutor Samuel Koenig, but in the end several prominent scientists came to her defense, and she was elected to the Academy of Sciences of the Institute of Bologna in 1746. Two years later, the Italian mathematician Maria Gaetana Agnesi (1718–1799) published one of the first and most complete textbooks on finite and infinitesimal analysis: *Instituzioni Analitiche (Analytical Institutions)*. The book was widely translated, and in 1750 Pope Benedict XIV appointed her to the chair of mathematics and natural philosophy at the University of Bologna, making her the first woman in history ever to be appointed to a mathematics university professorship.

Jumping ahead to 1890, we arrive in Cambridge, England, where Philippa Fawcett (1868–1948) has just become the first woman to earn a top score on the annual Mathematical Tripos exam (beating out all her male classmates), and students Isabel Maddison (1869–1950) and Grace Chisholm (1868–1944) are studying calculus and linear

Above: Cambridge University's Girton College in the 1890s. Mathematicians Grace Chisolm and Isabel Maddison were attending this all–women's college when they outperformed most male students on the 1892 Mathematical Tripos exam.

algebra at Girton College. Two years later, Maddison and Chisholm passed the Tripos exam, earning the equivalent of first-class degrees in mathematics from Cambridge. (Women were not granted formal degrees at the time.) They also challenged each other to complete the mathematics exam for the Final Honors School at Oxford, and this time Chisholm outperformed all other entrants.

Despite beating out male students at two of the world's most prestigious universities, women were not allowed to matriculate in graduate programs in England, so both Chisholm and Maddison ended up at the University of Göttingen in Germany, studying with group-theory pioneer Felix Klein. At this global hub of mathematics research—where Sofia Kovalevskaya earned her doctorate in 1874 (see page 26) and where the eminent mathematician Emmy Noether later helped establish the mathematics behind Einstein's theory of relativity (see page 36)—Grace Chisholm graduated magna cum laude to become only the second woman to be granted a PhD in Germany with her 1895 thesis "Algebraic Groups of Spherical Trigonometry." Her friend Mary Winston Newson (1869–1959), an American mathematician who arrived in Göttingen at the same time to study with Felix Klein, became the first American woman to earn a PhD in mathematics from a European university in 1897. As for Maddison, she specialized in differential equations and was awarded her PhD in 1896 from Bryn Mawr College in Pennsylvania. Fawcett's achievement led to her professorship at Cambridge's Newham College and inspired an anonymous poem:

Hail the triumph of the corset
Hail the fair Philippa Fawcett
Victress in the fray
    Crown her queen of Hydrostatics
    And the other Mathematics
Wreathe her brow with bay.

If you entertain objections
To such things as conic sections
Put them out of sight
    Rather sing of the essential
    Beauty of the Differential
Calculus tonight.

Worthy of our approbation
She who works out an equation
By whatever ruse
    Brighter than the Rose of Sharon

Are the beauties of the square on
The hypotenuse.

Curve and angle let her con and
Parallelopipedon and
Parallelogram
    Few can equal, none can beat her
    At eliminating theta
By the river Cam.

May she increase in knowledge daily
Till the great Professor Cayley
Owns himself surpassed
    Till the great Professor Salmon
    Votes his own achievements
      gammon
And admires aghast.[1]

In the intervening period between Agnesi's 1750 Bologna professorship appointment and Fawcett's triumph at the 1890 Cambridge Mathematical Tripos exam, several outstanding women who made pioneering advances in the field of mathematics were born. They come from China, Russia, France, Germany, and the United States, forming a global coalition of female mathematicians that has paved the way for countless women in STEM.

Left: Tripos top scorer Philippa Fawcett in her room at Newnham College, 1891.

WANG ZHENYI

1768 – 1797

*Qing Dynasty Astronomer and Mathematician Who
Explained Both Lunar and Solar Eclipses*

"There were times that I had to put down the pen and sighed.
But I love the subject, I do not give up."[1]

—Wang Zhenyi

Wang Zhenyi accomplished so much in such a short time on Earth—a mere twenty-nine years—it is fitting that she is remembered beyond it. There is a crater named for her on Venus, a nod to her expertise in astronomy. She is renowned for her abilities in the fields of poetry and mathematics as well.

Born during the Qing dynasty, Wang grew up in the Chinese province of Anhui nurtured by a family of scholars. Her father, Wang Xichen, a doctor who published four volumes about medicine called *Yifang yanchao* (*Collection of Medical Prescriptions*), taught her about medicine, geography, and mathematics. Her grandmother, Dong, gave her poetry lessons, and she learned astronomy from her grandfather, a former governor.

Wang spent a lot of time in her grandfather's large library poring over its many volumes and establishing a basis for her rich and varied education. When her grandfather died in 1782, the family moved to Jilin, near a portion of the Great Wall. During the five years they spent there, Wang became a well-rounded student with the assistance of her family and the wife of a Mongolian general named Aa, who taught her archery, martial arts, and equestrian skills. In her mid-teens, she traveled extensively with her father and befriended female scholars of modern-day Nanjing; these relationships enhanced her study of poetry.

On her own, Wang studied astronomy and mathematics, feeling thwarted at times by the difficulty of texts written in obscure, aristocratic language. She once wrote, "There were times that I had to put down the pen and sighed. But I love the subject, I do not give up."[2] This frustrating experience made her aware of how important it was for scientific texts to be clear and accessible to everyone, not just aristocrats. With this in mind, she rewrote mathematician Mei Wending's *Principles of Calculation*, calling her simpler version *The Musts of Calculation*. At age twenty-four, she devised a simpler method of performing multiplication and division and created the five-volume guide *The Simple Calculation Principles*. Wang also wrote the article "The Explanation of the Pythagorean Theorem and Trigonometry" as well as a paper about gravity that describes why people don't fall off the Earth even though it is a sphere.

Wang's desire to educate, and in particular to simplify complicated material extended to her astronomy studies, where she explained how equinoxes move and how to calculate this movement. In addition to

Below: Based on the original drawing by James Ferguson from his book Astronomy: Explained upon Sir Isaac Newton's Principles *published in 1756, this engraving by J. Mynde depicts the causes and effects of solar and lunar eclipses. Even before Ferguson published his famous work, Wang Zhenyi had composed the "The Explanation of a Lunar Eclipse" and developed a physical model to help explain the process.*

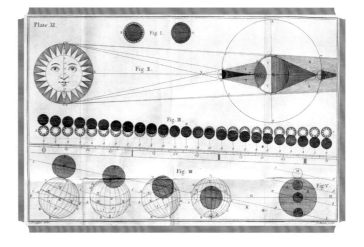

Opposite: The Porcelain Tower of Nanjing in China; the city has a long-standing history as a cultural and intellectual mecca, including during the time of Wang Zhenyi. Her time in Nanjing with other female scholars made a profound impact on her life and studies.

THE WANG ZHENYI CRATER:
TELL ME MORE

In 1994, scientists in the International Astronomical Union based in Paris, France, named a crater on Venus after the brilliant Wang Zhenyi. She joins the prominent list of women—including Jane Austen, Nellie Bly, Simone de Beauvoir, Dorothea Dix, and Margaret Sanger—who have been given a crater on the second planet from the sun. The diameter of the Wang Zhenyi crater (above) is 23.7 kilometers (14.7 miles), and it is located on Venus at latitude 13.2° and longitude 217.7°.[3]

analyzing the movement of the moon, she sought to remove some of the mystery of lunar and solar eclipses. In the eighteenth century, there were many legends and myths associated with eclipses; one legend described them as a sign of angry gods. "Actually, it's definitely because of the moon," Wang wrote of eclipses in one of her books, and she set out to demonstrate the phenomenon in a way that people could understand. Her exhibit of an eclipse, set up in a garden, was made up of a round table signifying the Rarth, a lamp for the sun, and a round mirror for the moon. By moving the objects, she showed how a lunar eclipse occurs as the moon passes into the Earth's shadow, making its path clear even for the youngest observers. Wang's related article—"The Explanation of a Lunar Eclipse"—remains highly accurate for its time.

Wang turned her travel experiences—informed by history and the classics—into "ci," or poetry, commenting on social issues such as the gap between poor and rich and the importance of giving equal opportunities to women and men. And she did so in a direct, unembellished style. One famous scholar remarked that her works "had the flavor of a great pen, not of a female poet."[4]

Wang married Zhan Mei Xuan Cheng at age twenty-five and continued to work as a poet, mathematician, and astronomer until her death at age twenty-nine. Her contributions have endured, and in 1994, the International Astronomical Union celebrated the naming of the Wang Zhenyi crater on Venus.

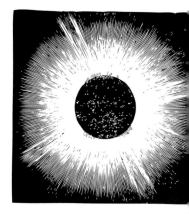

Above: A classic engraved illustration of the solar corona—that aura or glow visible around the sun, most visible during an eclipse—captured during the solar eclipse of 1860.

A RENAISSANCE WOMAN

In addition to her advances in both astronomy and math, Wang Zhenyi was an accomplished poet who was well ahead of her time thematically, as evidenced by some of the pieces she wrote during the Neo-Confucian Qing Dynasty:

It's made to believe,
Women are the same as Men;
Are you not convinced,
Daughters can also be heroic?[5]

SOPHIE GERMAIN

April 1, 1776 – June 27, 1831

A Revolutionary Mathematician

"How can I describe my astonishment and admiration . . . when a woman, because of her sex, our customs and prejudices, encounters infinitely more obstacles than men in familiarizing herself with [number theory's] knotty problems, yet overcomes these fetters and penetrates that which is most hidden, she doubtless has the most noble courage, extraordinary talent, and superior genius."[1]

—Carl Friedrich Gauss, ca. 1807

Above: *Death of Archimedes (1815) by French painter Thomas Degeorge.*

While the turmoil of the French Revolution raged, a precocious teen, confined to her home for safety, pulled books from her family's well-stocked library shelves and immersed herself in them. She taught herself Latin and Greek; she soaked up politics and philosophy. But when she read the story of the Greek mathematician Archimedes being killed by a Roman soldier because he refused to leave his mathematical diagrams, she became particularly intrigued with the power and possibilities of mathematics.

That teenage girl was Sophie Germain, born in Paris, France, the middle child of Marie-Madeleine Gruguelin and silk merchant Ambrose-François. Her parents' social standing placed them squarely in the bourgeois class. They lived comfortably and enjoyed the acquaintance of politically active and influential persons of the time. But being a woman during the eighteenth and nineteenth centuries had its disadvantages, of course.

While her parents probably appreciated that she was keeping herself occupied, they were not thrilled that she was studying a field considered inappropriate for women. Still, Germain persisted in her pursuit of mathematics, and with such passion that her parents confiscated her light, heat, and clothes to thwart her nighttime studies of English physicist Isaac Newton and Swiss mathematician Leonhard Euler. Somehow she found enough candles and quilts to persevere through nights so cold that ink froze in its well. Her parents eventually relented, realizing her passion for mathematics could not be extinguished.

Above: An 1889 engraving of Germain's mentor Joseph-Louis Lagrange.

Germain continued her studies without the aid of a tutor, grasping the fundamentals of differential calculus, among other topics. At age eighteen, she befriended students who attended the École Polytechnique, an academy for male mathematicians and scientists, and obtained notes from lectures of the many prominent mathematicians of the day. The work of one professor, Joseph-Louis Lagrange, particularly interested her. Taking the name of a deceased student as a pseudonym, Germain submitted a paper on mathematical analysis to Lagrange. Impressed with the insight displayed in the work, Lagrange sought out its author and was surprised to discover that "M. LeBlanc" was a woman. Nevertheless, he became Germain's mentor.

WHAT IS FERMAT'S LAST THEOREM?

Devised by Pierre de Fermat in 1637, this conjecture states that, while there are an infinite number of solutions for $n = 1$ and $n = 2$ (as exemplified by the Pythagorean triple*), no three positive integers x, y, and z satisfy the equation $x^n + y^n = z^n$ for any integer value of $n > 2$. This theorem can be divided into two cases—one involving all integers n that do not divide any of x, y, or z, and one involving integers n that divide at least one of x, y, or z. Germain's work proved the first case of Fermat's Last Theorem for all prime exponents less than 100 with the following statement:

Let p be an odd prime integer.
If there exists an auxiliary prime $P = 2Np + 1$
(where N is any positive integer not divisible by 3) such that:

1. if $x^p + y^p + z^p = 0$ (mod P^2), then P divides xyz, and
2. p is not a pth power residue (mod P).

Then the first case of Fermat's Last Theorem holds true for p.[3]

While Sophie Germain's Theorem would mark a significant advance toward the solution of Fermat's Last Theorem, the latter would remain unsolved for three and a half centuries.[4]

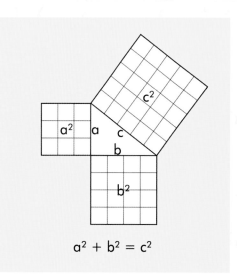

$$a^2 + b^2 = c^2$$

*The Pythagorean triple is any set of three integers n that satisfies the equation $a^2 + b^2 = c^2$ (the Pythagorean theorem). A triangle with three sides whose lengths correspond to the Pythagorean triple is known as a Pythagorean triangle, the most familiar of which is the 3-4-5 right triangle ($3^2 + 4^2 = 5^2$).

Lagrange introduced Germain to scientists and mathematicians she could not have interacted with otherwise. She collaborated with French mathematician Adrien-Marie Legendre, who later credited her in *Théorie des Nombres* for what is now known as Sophie Germain's Theorem, the first general result toward a proof of Fermat's Last Theorem.

Germain also corresponded with Carl Friedrich Gauss, a German mathematician known for his monumental 1801 work in number-theory, *Disquisitiones Arithmeticae*. Beginning in 1804, she wrote letters to him,

Above: An 1840 painting of Carl Friedrich Gauss by Danish artist Christian Albrecht Jensen.

again using the pseudonym M. LeBlanc. Gauss praised her number-theory proofs and shared them with his colleagues, believing them to be the work of a male mathematician. However, during the French occupation of his hometown of Braunschweig, Gauss learned the true identity of Germain when she used her family's influence to ensure his safety. He was grateful for the intervention and remained complimentary of her work.

In 1808, Germain became fascinated with the work of a German physicist and musician studying the vibration of elastic surfaces using a peculiar demonstration. After scattering fine sand on a metal plate, Ernst F. F. Chladni drew a violin bow across the plate's edge until a resonant vibration caused the sand to arrange itself in patterns. When the French Academy of Sciences introduced a contest seeking a mathematical explanation for these patterns, Germain accepted the challenge. (In fact, she was the only one to accept it.) After the conclusion of the two-year contest in 1811, Germain submitted her explanation anonymously. Her lack of formal education showed— she had not derived her hypothesis from principles of physics, which involves knowledge of the calculus of variations.

Above: An 1876 illustration of Chialdni's vibrating plates experiments from Scribner's monthly magazine **The Century.**

The contest was extended, and Germain got two more tries. On her next attempt—hers was, once again, the only entry—she received an honorable mention. For the final contest in 1815, Germain received a gold medal for her paper "Memoir on the Vibrations of Elastic Plates." Judges noted that her work was deficient in some areas. These shortcomings, however, were not addressed by other mathematicians for decades.

Above: An illustration of different patterns generated by Chialdni's experiment from **Popular Science Monthly,** *1873.*

A full seven years after Germain became the first woman to win a prize from the French Academy of Sciences, the academy secretary Jean-Baptiste-Joseph Fourier permitted her to attend their exclusive sessions. She also had the high honor of attending sessions of the Institut de France. With the help of many well-known male mathematicians of the 1820s, she refined her proofs as she continued to study the theory of elasticity and publish works that made significant contributions to the field.

Despite her dedication and talent, Germain did not receive the appreciation and respect that were her due. Both publicly and professionally, the mathematics community largely ignored her. She had made the first attempts to answer a problem so difficult it remained unsolved for 358 years, yet her work was routinely dismissed.

Above: Charles Marville's 1854 photograph of the River Seine and the Institut de France (domed building), near the Rue de Savoie, where Germain drew her last breath.

Germain persisted in her quest to uncover, in Gauss's words, "that which is most hidden." Supported by her father throughout her life, she continued work in mathematics as well as philosophy and psychology, completing papers on number theory and on the curvature of surfaces until she succumbed to breast cancer in June 1831 in the Monnaie quarter of Paris, where a plaque now commemorates her brilliant but brief life. Nearly two centuries after her death, her legacy includes an annual prize awarded by the French Academy of Sciences for advances in the foundations of mathematics, as well as several number-theory concepts named in her honor.

Opposite: This painting by French artist Henri Testiel shows members of the French Academy of Sciences and the institution's founder Louis XIV in 1667.

SOPHIE'S LEGACY

Germain curvature, n.:
a measurement of curvature defined by $(k_1 + k_2)/2$ when k_1 and k_2 are the maximum and minimum values of the normal curvature.

Sophie Germain prime, n.:
a prime p such that $2p + 1$ is also prime.

Sophie Germain's Identity[5], n.:
for any $\{x, y\}$, $x^4 + 4y^4 = ((x + y)^2 + y^2)((x - y)^2 + y^2) = (x^2 + 2xy + 2y^2)(x^2 - 2xy + 2y^2)$.

Sophie Germain's Theorem, n.:
(see page 22)

SOFIA KOVALEVSKAYA

January 15, 1850 – February 10, 1891

The First Female Mathematics PhD in Europe

"Many who have never had an opportunity of knowing any more about mathematics confound it with arithmetic, and consider it an arid science. In reality, however, it is a science which requires a great amount of imagination."[1]

—Sofia Kovalevskaya, 1895

More than one hundred years after her untimely death, Sofia Kovalevskaya[2] continues to be celebrated for her mathematical achievements, not only in her native Russia, but around the world. A €1.65 million prize is awarded annually in her name by the Alexander von Humboldt Foundation, and a 2009 short story by Nobel Prize–winning author Alice Munro was based on the nineteenth-century mathematician.

Left: Bust of Sofia Kovalevskaya by Finnish sculptor Walter Runeberg.

The middle child of Vasily Korvin-Krukovsky, an artillery general, and Yelizaveta Shubert, Kovalevskaya benefited from her family's noble social status and high level of education. She and her siblings were cared for by governesses and educated by tutors; the family's social circle included Russian novelist Fyodor Dostoevsky, whose proposal to Kovalevskaya's older sister, Anna, was rejected. An uncle is said to have sparked Kovalevskaya's curiosity in mathematics, discussing mathematical concepts with her when she was quite young. In her memoir, she wrote that their discussions instilled "in me a reverence for mathematics as an exalted and mysterious science which opens up to its initiates a new world of wonders, inaccessible to ordinary mortals."[3]

Above: Kovalevskaya's birthplace of Moscow is depicted in this 1878 painting by Vasily Polenov.

When Kovalevskaya began lessons with the family tutor, her love of mathematics became so obsessive that she started neglecting her other studies. Although her father ended her mathematics lessons, Kovalevskaya simply read a borrowed algebra book while everyone was asleep. Eager to expand her knowledge, she studied a physics textbook authored by a neighboring professor. The neighbor, impressed with her attempt at explaining concepts to him, encouraged her father to permit her to continue her studies in St. Petersburg.

Those mathematical concepts further inspired the young Kovalevskaya when, because of a wallpaper shortage, her nursery walls were covered with her father's notes on differential and integral analysis. This clever method of recycling lessons from Mikhail Vasilyevich Ostrogradsky, a leading mathematician of Imperial Russia, introduced the young girl to calculus. She was intrigued.

Below: The eminent Ukranian mathematician Mikhail Ostrogradsky, whose lecture notes covered the wall of young Sofia's bedroom in the 1860s.

After finishing her secondary education, Kovalevskaya sought a way to continue her studies at a university. The closest ones open to women were in Switzerland. She was young and unmarried; social custom did not permit women to travel alone nor live apart from their families without written permission of a father or husband. So, for convenience's sake, she married Vladimir Kovalevskaya, a paleontologist, and eventually moved to Heidelburg, Germany. She found a way around the restrictions the University of Heidelburg had on women attending classes by

THE CAUCHY-KOVALEVSKAYA THEOREM

Augustin Cauchy was a French mathematician who made ground-breaking contributions to the complex analysis of calculus and algebra. His 1842 findings and those Kovalevskaya made in 1875 were developed into the Cauchy-Kovalevskaya theorem, which proved a system of differential equations and led the way to modern mathematics.

obtaining permission from lecturers to unofficially attend their lectures. Her three semesters there established her as an uncommonly gifted student.

In 1870 at the University of Berlin, Kovalevskaya began her studies under Karl Weierstrass, one of the most renowned mathematicians of the time. He did not immediately take her seriously, but she proved herself by performing well on one of his problem sets. Since the university did not permit attendance by women, Weierstrass tutored her privately for four years. These productive years resulted in three papers—on partial differential equations, Abelian integrals, and Saturn's rings—which contributed to Kovalevskaya receiving her doctorate, summa cum laude, from the University of Göttingen in 1874. Her paper on partial differential equations was published in *Crelle's Journal* in 1875, an honor for such a young mathematician.

Unable to find worthy employment, despite having a PhD in mathematics and a recommendation from Weierstrass, Kovalevskaya and her husband returned to live in her hometown in Russia. There she wrote articles on various topics for a newspaper, and, in 1878, gave birth to a daughter.

When Kovalevskaya returned to mathematics in 1880, she did so with renewed vigor. She presented her paper on Abelian integrals at a scientific conference to great reviews and returned to Berlin to meet with Weierstrass. Kovalevskaya began studying the refraction of light in crystals. While she was away, tragic news reached her: her husband, whom she had already been separated from for a couple of years, had committed suicide in response to failed business ventures.

In 1883 Kovalevskaya eventually found work—thanks to Magnus Gösta Mittag-Leffler, a former student of Weierstrass—as a lecturer at the University of Stockholm. She was a popular lecturer; within a year her temporary position turned into a tenured professorship. During her five-year tenure, she was appointed editor of the mathematics journal *Acta Mathematica*, published her first paper on crystals, and was appointed chair of the Mechanics Department, becoming the first woman to hold a chair at a European university.

Despite her many accomplishments, Kovalevskaya's personal life was rocky at times. She battled depression

Above: Kovalevskaya's mentor Karl Weierstrass by German painter Conrad Fehr, 1895.

Below: Winter Scene of Stockholm (1899) by Swedish painter Alfred Bergström.

over the course of her life, and in 1887—four years following the suicide of her husband—her sister Anna succumbed to a lung infection, throwing Kovalevskaya into grief. While at her sister's deathbed, she conceived a story that became the basis of a play titled *The Struggle for Happiness*, which she cowrote with the Swedish actress and feminist Anne Leffler. Shortly thereafter, the Russian lawyer Maxim Kovalevsky arrived in Stockholm to give a series of lectures, and he and Kovalevskaya began a passionate affair. However, it fizzled when he asked her to give up mathematics to be his wife.

Kovalevskaya immersed herself in her mathematical research as she struggled to take her mind off her grief, disappointment, and heartache. In 1888, she submitted her anonymous paper "On the Rotation of a Solid Body about a Fixed Point" to the French Academy of Sciences as part of their Prix Bordin contest. Not only did she win, but judges deemed it so brilliant that they increased the prize money from 3,000 to 5,000 francs. Kovalevskaya continued research in this area and won a prize from the Swedish Academy of Sciences in 1889. She was also elected as the first woman to be a corresponding member of the Russian Imperial Academy of Sciences.

Kovalevskaya was at the height of her career when she succumbed in 1891 to pneumonia. Shortly before her death, she completed work on an autobiographical novel, *Nihilist Girl,* and an autobiography, *Recollections of Childhood*, which contains the following passage:

> It seems to me that the poet must see what others do not see, must see more deeply than other people. And the mathematician must do the same. [4]

Opposite: An anonymous painting of Sofia Kovalevskaya, ca. 1880.

WINIFRED EDGERTON MERRILL

September 24, 1862 – September 6, 1951

The First American Woman to Obtain a PhD in Mathematics

"She opened the door."

— inscription on a portrait of Winifred Edgerton Merrill presently on display at Columbia University

Above: A photograph of Columbia University and the Hudson River in New York published around 1903. Columbia was the first university in America to award a PhD to a woman—Winifred Edgerton—in 1886.

At the 1886 Columbia Commencement, Winifred Edgerton stood listening to the thunderous applause that surrounded her. During those incredible two minutes,[1,2] she thought about what it had taken to arrive at this moment: earning the first mathematics bachelor's degree from Wellesley College in 1883; petitioning to take part in the PhD program in mathematics and astronomy at Columbia University and, at first, being denied; pleading her case with Columbia trustees; pursuing her studies virtually alone; completing an original thesis on integrals. And now here she was, the first American woman to be awarded a PhD in mathematics and the first woman to graduate from Columbia University.

Winifred Edgerton Merrill's journey was marked by diligence and talent. She was born Winifred Edgerton in Ripon, Wisconsin, and received her early education from private tutors before earning her BA (at age sixteen) with honors in mathematics from Wellesley College, one of the first American colleges for women. She taught briefly, studied the orbit of the Pons–Brooks comet at the Harvard College Observatory, then—desiring access to a high-powered telescope to continue her studies—applied to study at Columbia University in the male-dominated field of mathematics and astronomy.

With her solid background in astronomy from her Wellesley years, Edgerton thrived in her doctoral program, completing it in just over two years. She did so in solitary fashion, as the administrators did not allow her to interact with her male counterparts nor attend lectures. Instead she studied from the course text on her own. During long nights at the telescope, she coped with loneliness by setting out her doll collection to keep her company, quickly putting them away when anyone approached.[3]

Her original two-part thesis (the first in mathematical astronomy), included the orbit computation of the 1883 comet (the Pons-Brooks Comet, see page 34) which was only the second time the comet had been observed. The second part of her thesis was in pure mathematics and was titled "Multiple Integrals (1) Their Geometrical Interpretation in Cartesian Geometry, in Trilinears and Triplanars, in Tangentials, in Quaternions, and in Modern Geometry; (2) Their Analytical Interpretations in the Theory of Equations, Using Determinants, Invariants and Covariants as Instruments in the Investigation."[4]

The thesis explored infinitesimals in various systems from analytical geometry, including Cartesian, oblique, polar, trilinear, triplanar, tangential, and quaternions. Edgerton began by using a geometric approach to find the infinitesimals for length, area, and volume for these various systems. She stated that the presentation of the trilinear and triplanar coordinate systems was new. Then she presented the transformation

NEWSWORTHY?

When Winifred graduated from Columbia in 1886, the *New York Times* was there to report on the momentous event, including her fashion choice. According to their report, "She was modestly dressed in a walking dress of dark brown stuff, trimmed with velvet of the same material, and wore a brown chip hat which had a pompon of white lace and feathers."[5]

Below: The grounds of the Harvard College Observatory circa 1899. Merrill used the high-powered telescope called the "Great Refractor," which had been installed in the Observatory in 1847. It was the largest telescope in the United States for twenty years.

THE PONS-BROOKS COMET

French astronomer Jean-Louis Pons was a concierge at the Observatory of Marseilles in France when he first discovered the "nebulous" comet that seemed to have no definable tail on July 21, 1812. He wouldn't report his findings until the next day, though, July 22. The comet's orbit has been estimated at 70.68 years, and it was spotted in Phelps, New York, next time by William Robert Brooks on September 1, 1883. Both Brooks and Pons discovered numerous comets in their lifetimes, but it was Pons who smashed the world record with at least thirty-seven discoveries in his lifetime. The Pons-Brooks comet can be seen with the naked eye at points during its orbit, and it belongs to the Halley's family of comets. It is expected to make its next approach to Earth in the year 2024.[6,7]

Above: A portrait of Winifred Edgerton Merrill

method abridged from Bartholomew Price's book on differential and integral calculus for transformation of multiple integrals from one system to another. Part of her original work thus included using this method for transformations from the Cartesian system to that of triplanars and tangential coordinates. Edgerton then examined the infinitesimals for area and volume obtained by the geometric approach as well as the analytical method for various systems and showed their equivalence. In addition, new work was done with transformations and the quaternion system. One of the new results in her thesis involved obtaining relations between the Cartesian and oblique systems and the oblique and triplanar systems to obtain the needed equation arrays for the Cartesian and triplanar systems.[8]

After marrying Columbia School of Mines graduate Frederick James Hamilton Merrill in 1887, Winifred Edgerton Merrill was one of five people in 1889 who helped found Barnard College, New York's first secular institution to award the liberal-arts degree to women, though she was ultimately forced to resign from the all-male group: "We had our meetings downtown. My husband objected very much. He thought it was entirely improper for me to go to a man's office downtown. I had to resign from the Committee."[9]

Merrill taught mathematics at several institutions, and, in 1906, she founded the Oaksmere School for Girls, a college-preparatory school that remained in operation for twenty-two years. She and her husband had four children. An advocate for education and for women, she wrote articles and spoke widely before her death in Fairfield, Connecticut, in 1951.

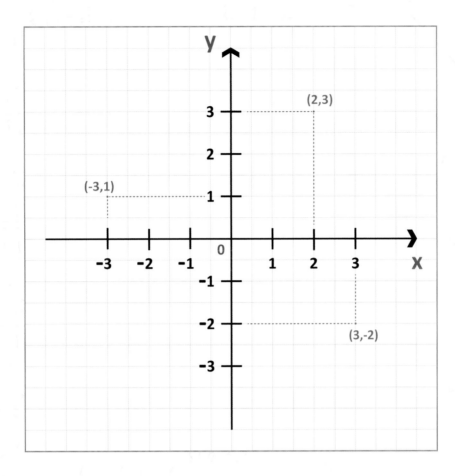

WHAT IS A
CARTESIAN COORDINATE SYSTEM?

Developed by René Descartes in the seventeenth century, the Cartesian Coordinate System is a useful method for identifying the position of any given point between two perpendicular lines. In the illustration above, for example, the coordinates for the point in the upper-right quadrant are identified as 2,3. This means you have to move two numbers horizontally along the x-axis and three up on the vertical y-axis to reveal the point's position. Think of it like a game of Battleship!

EMMY NOETHER

March 23, 1882 – April 14, 1935

Influencing Einstein

"Noether was the most significant creative mathematical genius thus far produced since the higher education of women began."[1]

—Albert Einstein, 1935

It's not easy to overstate the brilliance of Amalie "Emmy" Noether (pronounced NER-ter). The eldest child of a prominent mathematician and professor, she did far more than carry on her family's academic tradition. Not only did she make groundbreaking developments in abstract algebra,[2] but history's most iconic genius relied on her mathematical expertise as he devised his revolutionary theory on the nature of gravity, space, and time.

Left: Albert Einstein, 1916.

Noether grew up in a well-to-do household in the Bavarian town of Erlangen, Germany. Her mother, Ida Amalia Kaufmann, was the daughter of a wealthy Jewish merchant family from Cologne, and her father, Max Noether, an innovative leader of nineteenth-century algebraic geometry. He was well published and, as a distinguished professor at both the University of Heidelberg and the University of Erlangen, was known for being a patient and encouraging teacher.

Love of science and mathematics permeated the household. The Noether children grew up in a home filled with books, rich discussions, and the children of their father's academic friends. Noether's brother Alfred studied chemistry, while Fritz shared his father's enthusiasm for mathematics.

Above: This 1916 postcard shows the University of Erlangen, where Noether studied and taught mathematics (without pay) in the early 1900s.

As a girl, Noether did not have the luxury of studying anything she wanted like her brothers did. Instead she learned skills that were traditional for young women at that time: cooking, cleaning, and playing the clavier, an early keyboard instrument. By most accounts she was clever, animated, and well-liked by her peers, despite having a slight lisp. She enjoyed dancing, studied French and English, and passed exams in 1900 that would allow her to teach at schools for young women. But Noether had other plans.

At age eighteen, Noether got permission to audit classes from the German professors at University of Erlangen, where one of her brothers was a student and her father was a professor. This was the only way women could study at German universities. After sitting in on classes for two years (one of only two women doing so), she passed a matriculation exam in 1903. She also spent time at the University of Göttingen, attending lectures by noted mathematicians of the time, including David Hilbert and Felix Klein, two men with whom she would later collaborate. Her father's tutelage and her deep study of the field provided Noether with a strong background for a career in mathematics.

By 1904, women were allowed to matriculate at German universities, and Noether returned to the University of Erlangen. She

Above: In this postcard from April 10, 1915, Noether discusses algebraic concepts with her Erlangen colleague Ernst Fischer.

Below: David Hilbert, shown in this 1907 photograph, was one of Noether's biggest supporters at the University of Göttingen, where the idea of a female professor left many colleagues aghast.

studied invariant theory and algebraic geometry under Paul Gordan and was awarded a PhD in mathematics, summa cum laude, from Erlangen in 1907. Her dissertation, "On Complete Systems of Invariants for Ternary Biquadratic Forms," was well received by the mathematics community, but she was not particularly impressed with her own work, describing it as "crap."[3]

While rules regarding university entrance had changed, longstanding biases about women teaching at these same universities had not. Noether was well qualified to teach at the University of Erlangen, but was not allowed to do so. She stayed on, however, making a name for herself at the Mathematical Institute of Erlangen, where she conducted research on theoretical algebra. She also helped her father by filling in to teach his classes when he could not and helping advise several of his doctoral students (without pay, of course). She was interested in her father's work and generalized some of his theorems.

As her publications began circulating, Noether's reputation grew. In 1908 she was elected to the Circolo Matematico di Palermo. Then, in 1909, she was invited to become a member of the Deutsche Mathematiker-Vereinigung. She began lecturing at mathematics gatherings throughout Europe. In 1915 two of her colleagues, Felix Klein and David Hilbert, were working at the University of Göttingen on Einstein's general relativity theory and realized they needed her expertise in invariant theory.

Noether accepted the invitation from Klein and Hilbert and found herself back at Göttingen, this time as an unpaid lecturer, advertising her courses under Hilbert's name. Meanwhile, she became an invaluable resource to Albert Einstein, formulating what became known as Noether's Theorem. In essence, this finding states that wherever there is symmetry in nature, there is a corresponding conservation of energy, momentum, and electric charge. For example, a symmetry of time (e.g., no matter when you toss a ball in the air, the trajectory remains the same) can be explained by the notion of conservation of energy (i.e., the total energy of an isolated system is invariant, or "conserved," over time).[4]

Noether's mathematical proof of the relationship between time and energy in the universe was later described as "certainly one of the most important mathematical theorems ever proved in guiding

WHAT IS INVARIANT THEORY?

This branch of abstract algebra examines the actions of groups (i.e., a set of elements, such as integers, with an operation, like addition, that combines two elements to form a third) on invariant vector spaces or other algebraic varieties (i.e., a set of solutions of polynomial equations that don't change, no matter how the integers are manipulated). A real-world example of invariance is found when you examine the relationship between a star and one of its planets: while the shape and radius of the planetary orbit change with time, the gravitational attraction between planet and star is invariant.

the development of modern physics, possibly on a par with the Pythagorean theorem."[5]

Despite Noether's high esteem among the most brilliant physicists and mathematicians of the time, several Göttingen faculty members remained uncomfortable with the idea of having students learn "at the feet of a woman." In a famous retort, Hilbert replied, "We are a university, not a bath house."[6] With help from Hilbert and Klein, along with a change in attitudes brought about by the German Revolution following World War I, in 1919 Noether finally gained permission to obtain her habilitation—a postdoctoral qualification

Above: Emmy Noether, ca. 1930.

needed to lecture at a German university—and to be named a privatdozent.

While she was paid very little to teach, Noether was fortunate to have an inheritance from her family, which permitted her to concentrate on her students and research. During her time at the university, she was known as warm, caring, and tough. She often shared food and belongings with her students, encouraging them to develop her ideas (over which she claimed no ownership) and even inviting them to her apartment for discussions. Despite her a propensity to talk quickly—reports say she also gesticulated wildly during her lectures—she developed a loyal following known as "Noether's Boys," some of whom went on to become great mathematicians.

In addition to her eponymous theorem, Noether was integral to the development of modern abstract algebra, publishing her groundbreaking work *Idealtheorie in Ringbereichen* in 1921. This work gave rise to the term Noetherian and resulted in principles unifying topology, geometry, logic, and algebra. Her later work on non-commutative algebras[7], representation theory, hyper-complex numbers, and linear transformations earned her the 1932 Ackermann-Teubner Memorial Prize in Mathematics, which she received jointly with Emil Artin. Noether turned fifty in the same year and gave a lecture to

WHAT MAKES SOMETHING "NOETHERIAN"?

In mathematics, the term Noetherian is used to describe objects that satisfy an ascending or descending chain condition—that is, the sequence of ascending or descending elements eventually terminates (i.e., is not infinite). In algebraic terms, given any ascending sequence ($a_1 \leq a_2 \leq a_3 \leq \ldots$) or descending sequence ($\ldots \leq a_3 \leq a_2 \leq a_1$), there exists a positive integer n such that $a_n = a_n + 1 = a_n + 2 = \ldots$.[8] Examples of Noetherian mathematical objects include groups, rings, modules, relations, topological spaces, and schemes.

the International Congress of Mathematicians in Zurich, Switzerland, on hypercomplex number systems.

While Noether lived for mathematics, she couldn't ignore the events unfolding around her. Noether was Jewish, a pacifist, and social democrat. When Hitler and the Nazis came to power in Germany in 1933, they ordered the removal of Jewish professors from the universities. This affected both Noether and her brother Fritz, who also taught mathematics in Germany. Her colleague Hermann Weyl—an incredible mathematician and physicist in his own right—marveled at her grace during that time:

Above: Attendees at the 1932 International Congress of Mathematicians in Zurich included quantum physics pioneer Wolfgang Pauli and cybernetics founder Norbert Weiner, among many other top minds.

> *In the midst of the terrible struggle, destruction and upheaval that was going on around us in all factions, in a sea of hate and violence, of fear and desperation and dejection—you went your own way, pondering the challenges of mathematics with the same industriousness as before. When you were not allowed to use the institute's lecture halls you gathered your students in your own home. Even those in their brown shirts were welcome; never for a second did you doubt their integrity. Without regard for your own fate, openhearted and without fear, always conciliatory, you went your own way.*[9]

Fritz left for Moscow, and Emmy accepted a teaching position at Bryn Mawr, a women's college in Pennsylvania, for an annual salary of $4,000. For two years, students at Bryn Mawr and at nearby Princeton University (where Albert Einstein was teaching) sat in on the energetic, spontaneous, and unstructured lectures of this brilliant, boisterous mathematician who slipped them money when they were in need and took them on hikes in the wilderness. Sadly, in 1935 she passed away following surgery to remove a uterine tumor. At her memorial, Weyl shared some touching words about the incredible mind and spirit that was suddenly gone from the world:

Above: This 1931 painting by Albert Reich shows an advancing army of "Brownshirts" with Nazi flags as Hitler was consolidating power in Germany. Some of these Nazi Party soldiers were students of Noether, who welcomed them into her home nonetheless.

I have no reservations in calling you the greatest [woman mathematician] that history has known. Your work has changed the way we look at algebra . . . No-one, perhaps, contributed as much as you towards remolding the axiomatic approach into a powerful research instrument, instead of a mere aid in the logical elucidation of the foundations of mathematics, as it had previously been.[10]

Her brother Fritz, who was unable to attend the service, didn't fare much better: six years later, he was arrested in Tomsk as a "German spy" and sentenced to twenty-five years imprisonment during the Soviet Union's Great Purge. Four years into his sentence, he was executed.

Today, Noether is celebrated widely, with scholarships, lectureships, and even a crater on the moon named in her honor.

Opposite: The archery team at the all-female Bryn Mawr College, where Noether taught from 1933 to 1935, shows they are a force to be reckoned with. After her death, Noether's ashes were buried on the school's grounds.

Breaking the PhD Color Barrier

EUPHEMIA HAYNES

September 11, 1890 – July 25, 1980

"For a person of intelligence is well equipped to solve the problems of life . . . . Let each defeat be a source of a new endeavor and each victory the strengthening of our spirit of gratitude and charity towards the unsuccessful."

—Martha Euphemia Haynes's valedictory speech, 1907

Above: Dr. Euphemia L. Haynes, ca. 1907.

After the Civil War, Washington, D.C., became a hub for freed slaves and, thus, a home to a large and dynamic African American population. Among the notable nineteenth-century residents was the abolitionist, suffragist, writer, and statesman Frederick Douglass, who settled his family on a hill overlooking the Anacostia River in southeast Washington in 1877. Another well-to-do resident of mixed race was third-generation Washingtonian Lavinia Day, who taught kindergarten in the public schools and was active in the Catholic church. On October 30, 1889, she married William Lofton, a well-known dentist and civil-rights activist in the D.C. community. Almost a year later, on September 11, 1890, Martha Euphemia Lofton was born.

Although the couple's social standing afforded Euphemia opportunities that many other African American families lacked, by her seventh birthday her parents had split. She and her younger brother, Joseph, were raised by their mother and had little contact with their father. But young Lofton was determined to prove her worth. In 1907 she was named valedictorian at the M Street High School, and two years later she graduated with distinction from the Miner Normal School. At Smith College in Massachusetts she minored in psychology and majored in mathematics. She also began her career in the D.C. public school system, teaching elementary as well as high-school classes in English and mathematics. In 1917 she married her high-school sweetheart, Harold Appo Haynes, a fellow teacher who had earned an electrical engineering degree from the University of Pennsylvania. Together they attended the University of Chicago, where Euphemia Haynes took graduate-level mathematics courses.

Above: This 1910 photograph shows first graders at the Minor Normal School brushing their teeth. Lofton-Haynes attended the same school and graduated with distinction just a year earlier in 1909.

After receiving her master's degree in education from the University of Chicago, Haynes established and chaired the mathematics department at Miner's Teachers College, where she remained for almost three decades. While professor there, she pursued a doctorate in mathematics from the Catholic University of America. Under the advisement of Professor Aubrey Landry, who specialized in algebraic geometry, she completed the thesis "Determination of Sets of Independent Conditions Characterizing Certain Special Cases of Symmetric Correspondences," and in 1943 she became the first African American woman to earn a PhD in mathematics.

Like her mother, Haynes was a lifelong Catholic and dedicated herself to her community. She helped found the Catholic Interracial Council of the District of Columbia, was president of the Washington Archdiocesan Council on Catholic Women, and supported the Fides House, a neighborhood hospitality house organized by the Catholic

Right: The main building of the Catholic University of America, which became the first institute of higher learning to award a doctorate degree in mathematics to an African American woman in 1943.

University of America that provided services, assistance, and educational help to the poor. She was a member of the D.C. branch of the National Conference of Christians and Jews and a board member of Catholic Charities. In 1959—the same year she retired from Miner's College, which had by then had become the University of the District of Columbia—Pope John XXIII honored her with the papal medal "Pro Ecclesia et Pontifex" for her distinguished service to the church and her community.

Together with her husband, who served as principal and superintendent of D.C.'s segregated schools, Haynes was a staunch advocate for equal education. Some of her most notable work occurred during her time as a member of the District of Columbia Board of Education from 1960 to 1968, a tumultuous time in American history. As a board member—and, later, as the first female board president—Haynes advocated for poor students and better schools. The D.C. school system had long operated a "track system" that placed primarily African American and poor students in academic or vocational programs based on their abilities during the early grades. There was no opportunity to switch tracks, even if their achievements and interests changed. Subsequently, many of these students ended up in noeducational vocational programs. Haynes denounced this segregated system and was among those to celebrate its abolishment in June 1967. She also championed collective bargaining rights for public school teachers before retiring from board service in 1968.

The same year Haynes became the first female African American mathematics PhD, she spoke of the connection between the pursuits of mathematics and world peace:

Above: Pope John XXIII honored Haynes with the prestigious Papal medal (shown) in 1959.

> *In whatever corner of the world they may find themselves, mathematicians, like all scientists, are bonded together by a universal desire to understand life. Cooperation is natural, it is easy, it is necessary in the all-out effort of science to establish truth.[2]*

When Haynes passed away at the age of ninety, she bequeathed $700,000 to the Catholic University of America, which has cemented her legacy with the establishment of an endowed chair in the Department of Education, an annual colloquium, and a perpetual student loan fund— each bearing her name.

PART II

From Code Breaking to Rocket Science

T
he late nineteenth century saw the rise of female "computers," beginning with American physicist and astronomer Edward Charles Pickering's "harem," now popularly known as the Harvard Computers. In 1881, Pickering was director of the Harvard College Observatory when he became frustrated by his male assistants and decided to give their work to his maid, Williamina Fleming. She turned out to be much more efficient with every task, and five years later, he started building a small staff of skilled women to pore over astronomical data and classify stars. Among them was Wellesley graduate Annie Jump Cannon (1863–1941), who had studied alongside Winifred Edgerton Merrill (see page 32). Despite receiving wages of just 25 to 50 cents an hour—barely above those of unskilled factory workers—these women made possible the publication in 1890 of the first Henry Draper Catalog, which included more than ten thousand stars classified by their emission spectrums. Cannon developed the Harvard Classification Scheme, which served as the basis of today's stellar classification system. Fellow computer Henrietta Swan Leavitt (1868–1921) discovered a direct relationship between the brightness of Cepheid Variable stars and the length of their pulsation periods, which in turn became an essential tool in the measurement of the universe.

Opposite: Annie Easley

Above: A ca. 1890 photograph of the Harvard Computers, including Williamina Fleming (top center).

By the early twentieth century, the women's college Bryn Mawr and the coed University of Göttingen were two major hubs for women seeking advanced degrees in mathematics. However, as the United States and several major European powers were drawn into World Wars I and II, suddenly women trained in mathematics were in high demand for defense-related occupations, including cryptanalysis and ballistics. One such woman was Agnes Meyer (1889–1971), a polymath and polyglot from the Midwest who headed the mathematics department at Amarillo High School in Texas. Her proficiency in mathematics as well as German, French, Latin, and Japanese made her an excellent candidate for wartime cryptanalysis, so in 1918, she enlisted in the Naval Reserve as chief yeoman.

In a very short period, Agnes codeveloped the navy's Communications Machine ("CM"), which remained its standard cipher machine throughout the 1920s. She taught cryptology to both Laurance Safford, the first head of the navy's Cryptographic Research Desk, and the celebrated cryptanalyst Joseph Rochefort, who helped her break the Japanese naval Red Book code in 1926 and the Blue Book code in 1930. Thomas Dyer, who led the team that was responsible for most of the breakthroughs in reading Japanese naval communications during World War II, was also one of her pupils. In 1935, she was the first in a group of skilled analysts to decrypt the Japanese naval "Orange" M-1 cipher machine, and she went on to serve as "principal cryptanalyst" for the navy until 1950. Meyers's exceptional skill at deciphering military and diplomatic communication over her thirty-year career earned her the nicknames "Madame X" and "the first lady of cryptology."

Left: Agnes Meyer's decryption work was essential in the Allied victory over the Japanese in many naval battles during the Pacific War, including the decisive Battle of Midway in 1942 depicted in this painting.

In 1907—the year Agnes Meyer enrolled at Otterbein University in Columbus, Ohio, where she studied statistics, physics, and music alongside mathematics and languages—Irene Kaminka was born in Vienna, Austria, to Rabbi Armand Kaminka and his wife. After graduating from high school, Irene attended the Vienna University of Technology, where she undertook studies in descriptive and projective geometry. By 1941, she and her husband, Eric Fischer, had fled the Nazis with their young daughter, taking a boat around the southern tip of Africa and across the Atlantic to Boston, Massachusetts. As they established a new life in America, Fischer took a job grading blue books for the mathematician and former child prodigy Norbert Weiner at Massachusetts Institute of Technology. She also applied her projective geometry expertise to creating stereoscopic trajectories for Weiner's colleague John Rule before taking a position as mathematics instructor at Buckingham Browne & Nichols School in nearby Cambridge.

After World War II, the United States was investing heavily in several technologies—including computers, aviation, and satellite communication—aimed at giving the US and its allies an economic and military advantage over the Soviet Union and other nations. Enter Irene Fischer. In 1946, she took a job at the Army Map Service (AMS) in Potomac, Maryland, and began a twenty-five-year career developing what became the World Geodetic System that makes the Global Positioning System (GPS) possible. Fischer's geodetic calculations were also used to determine the accurate parallax of the moon, the lingering effects of the last ice age, and the oblateness of the Earth. She wrote an impressive 120 publications over her career,

Above: Thanks in part to mathematicians like Helen Fischer, we can locate and determine the time and path of navigation to any location in a matter of seconds.

Above: Kay McNulty and other hired "computers" calculate ballistics trajectories during World War II.

and, in 1967, she received the US Army's Distinguished Civilian Service Award.

A highly skilled mathematician, writer, and speaker, Fischer also joined the AMS at a fortuitous time, when computer and satellite technology allowed for significant breakthroughs in geospatial measurement. Some of these advances were made possible by the Irish American mathematician and computer programmer Kay McNulty (1921–2006), who led a team of women in developing the first general-purpose electronic digital computer. Completed in

Above: A woman with a microscope and Friden calculating machine collects and analyzes data at NASA's Langley Research Center in 1952.

Above: The work of many women in this book helped make countless discoveries about our Universe via their calculations for NASA, which launched the Voyager spacecraft in 1977. After forty years these satellites have captured thousands of images of our Solar System, including those in this montage of our eight planets and four Jovian moons, and are now headed into interstellar space.

1946—the same year Irene Fischer joined AMS—the Electronic Numerical Integrator and Computer (ENIAC) was dubbed a "Giant Brain" by the press and succeeded in calculating ballistics trajectories 2400 times faster than a human could. Like too many women in this book, however, the six original programmers of ENIAC (dubbed the "Refrigerator Ladies," since everyone assumed the women in the photographs were simply models advertising the new machine) were not recognized until decades later. In 2010, documentary filmmaker LeAnn Erickson chronicled their achievements in *Top Secret Rosies: The Female "Computers" of WWII.*

Alongside these incredible female mathematicians were countless others, making quiet contributions to the world economies and the American way of life. One of the most notable employers of female mathematicians was the National American Space Agency (NASA), a competitor in the international Space Race that sent men to the moon and satellites beyond our solar system. A diverse group representing multiple ethnicities and varied economic backgrounds, these women made space travel possible and allowed us to gather crucial data about the nature of the universe, as well as its stellar and terrestrial inhabitants. Here are their stories.

GRACE HOPPER

December 9, 1906 – January 1, 1992

Creator of the First Computer Compiler

" [Grace Hopper] emphasized the future because she feels that she actually lives in the future. This is only the beginning of the computer age . . . we are only beginning to know what to do with computers. "[1]

—Vassar College's *Miscellany News*, 1967

Grace Hopper's favorite saying, "Dare or do," continues to inspire as the largest gathering of women in technology meets annually for the Grace Hopper Celebration, established in 1994. More than eighteen thousand participants attended the 2017 event.

Left: Grace Hopper distributed thousands of these pieces of wire that represent the distance an electrical signal travels in a nanosecond (i.e., a billionth of a second).

Encouraged by her parents to value education, Hopper, born Grace Brewster Murray in New York City, was naturally curious and took apart alarm clocks and other gadgets for fun. Hopper received a bachelor's degree in mathematics and physics (graduating Phi Beta Kappa) from Vassar College in 1928, and in 1930 earned a master's degree in mathematics from Yale. She spent thirteen years teaching at Vassar, during which time she also earned her doctorate at Yale with a 1934 dissertation titled "New Types of Irreducibility Criteria." A paper published on Pythagorean number theory soon followed in *American Mathematical Monthly*.

Above: The entrance to Vassar College, where Hopper earned her bachelor's degree in physics and mathematics in 1928.

In 1943, during the Second World War, Hopper took a leave from Vassar to join the navy. Despite being considered at first too old and too slight in stature, she was eventually accepted into the US Naval Reserve to serve in WAVES (Women Accepted for Volunteer Emergency Service). She was assigned as a lieutenant to the Bureau of Ships Computation Project at Harvard University, where she worked on the IBM Automatic Sequence Controlled Calculator (ASCC), more commonly known as the Harvard Mark I, the world's first large-scale computer. Here she put her mathematical skills to good use to aid the war effort, especially in calibrating different types of artillery. That work reminded her of her days taking apart clocks, and she was eager to figure it out. She called the computer "an impressive beast"[2]—it was fifty-one feet long, eight feet high, and weighed about five tons. Hopper is known as the third computer programmer because she was the third person to master the Mark I, even writing its 561-page user manual. She also went on to become an expert with its successors, Mark II and Mark III. For her work on these computers, she received the Naval Ordnance Development Award in 1946.

In 1949, Hopper entered the business world, joining the Eckert-Mauchly Computer Corporation as senior mathematician. There she mastered yet another computer, the UNIVAC (Universal Automatic Computer) I, the first commercial computer. Instead of retyping the same commands to run certain programs, which easily introduced errors, Hopper suggested that the programmers write the commands down in a notebook to create "shared libraries of code."[3] This simple innovation led Hopper and her team of programmers to create A-O, the first compiler that translated mathematical code into binary code (the number system using only 0s and 1s), which remains the language of machines today.

From there, Hopper helped develop FLOW-MATIC, a word-based compiler that made UNIVAC I and II more accessible in the everyday workplace, programming twenty word-based statements. Despite being repeatedly told that computers don't understand words, Hopper worked for many years to champion the idea of an entire computer language based in English. In 1959, she helped develop and standardize the first universal computer language, COBOL (Common Business-Oriented Language), based on FLOW-MATIC. Through the combination of her knowledge, her passion for COBOL, and her gift for communicating, Hopper was able to influence the military, specifically the navy, and businesses to use COBOL, making it the "most extensively used computer language" by the 1970s.

Above: Hopper and her Cruft Research Lab colleagues in front of the Harvard Mark I computer during World War II.

Not only was Hopper active in the private sector and the navy—retiring in 1986 as a rear admiral—she also was involved in academics and teaching. Throughout her life she held numerous visiting positions at colleges and universities in addition to consulting and lecturing for the Naval Reserve. She continued to share her wisdom as a senior consultant to Digital Equipment Corporation after retirement.

Hopper received many awards and accolades for both her technological and military leadership, including the first Computer Science Man-of-the-Year Award from the Data Processing Management Association in 1969. The Department of Defense presented her with its highest award, the Defense Distinguished Service Medal. In 1973, she was not only the first woman, but also the first American, to be elected as a Distinguished Fellow of the British Computer Society. In 1987, the navy launched the USS *Hopper*, known as "Amazing Grace" by its crew, a high-tech missile destroyer. In 2016, nearly a quarter century after her death, Hopper received the Presidential Medal of Freedom. Despite her many technological breakthroughs, Hopper believed that her greatest achievement was "all the young people I've trained."[4]

Above: Hopper at her desk in the Harvard's Computation Laboratory, 1947.

Below: Hopper helped to develop UNIVAC 1, the first commercial computer. Shown here is the UNIVAC 1105 model, which replaced UNIVAC I at the US Census Bureau in the early 1950s.

THE ORIGINS
OF THE
COMPUTER "BUG"

One night, when there was a malfunction with the Mark II, Hopper and her colleagues discovered that the culprit was a large moth caught in the machine. From then on, she and her colleagues began calling any future computer problems that arose a "bug" and the fixing of said problems a "debugging."

Opposite: Hopper was appointed by former US president Ronald Reagan to the rank of commodore (now classified as "rear admiral") in 1983.

Below: The USS Hopper destroyer launching a missile on the Pacific Ocean in 2009.

MARY GOLDA ROSS

August 9, 1908 – April 29, 2008

Pioneering Native American Rocket Scientist

"To function efficiently in today's world, you need math.
The world is so technical, if you plan to work in it,
a math background will let you go farther and faster."[1]

—Mary Golda Ross

The Cherokee tradition has long advocated for equal education for boys and girls, so perhaps that is part of the reason Mary Golda Ross never cared that she was the only girl in a class full of boys. In fact, her mathematical prowess and staunch determination would one day help her shape the American space program.

Ross grew up in Park Hill, part of southwestern Cherokee County in Oklahoma. Chief John Ross, leader of the Cherokee nation from 1828 to 1866, was her great-great-grandfather. Her family had settled in Oklahoma by way of the Trail of Tears, the US government's mid-nineteenth-century mandated relocation of Cherokees from Georgia to Indian Territory. Ross was drawn to math at an early age; as she explains in an interview with Laurel M. Sheppard, "Math was more fun than anything else. It was always a game to me."[2] She credits her parents in part for the strength of her education. "My parents believed an education was necessary to make something of yourself," she said "I did not dare miss a day of school."[2]

At age sixteen, Ross entered the college her great-great-grandfather had helped found, Northeastern State Teacher's College (now Northeastern University), and graduated at age twenty with a degree in mathematics. For nine years, she taught math and science to children in rural Oklahoma public schools and also served as a girls' advisor at a coed Pueblo and Navajo school. By 1937, she decided she wanted to see more of the world. She was hired as a statistical clerk for the Bureau of Indian Affairs in Washington, D.C., and, while employed there, returned to school to attain a master's degree in mathematics in 1938 from Colorado State Teachers College (now the University of Northern Colorado at Greeley).

Above: Mary Golda Ross (second from the right) ca. 1950. She is with her family: Charles Ross and his wife Maxine Miller Ross (far left), Mary's mother (center) Mary Henrietta Moore Ross and sister (far right) Frances Curtis Ross Glidewell.

DID YOU KNOW?

Above: Celebrity panelists are blindfolded in this 1952 episode of What's My Line?

The popular TV show *What's My Line?* that ran from 1950 to 1967 featured celebrity guests who would try to guess the occupations of contestants. Mary Golda Ross appeared on the June 22, 1958, episode and stumped her celebrity inquisitors. None of the celebrities—including Jack Lemmon—could guess that she designed rocket missiles and satellites for Lockheed.

SKUNK WORKS:
THE BEGINNING

Above: Clarence "Kelly" Johnson in 1965 standing with scale models of his plane designs.

Started by the innovative Clarence "Kelly" Johnson in 1943 after he delivered a design proposal for the XP-80 Shooting Star jet fighter, Skunk Works became a small guerilla group within Lockheed Aircraft Corporation. Because they didn't follow the rules or organization of the larger business, the engineers were able to deliver designs and products faster and more efficiently. The XP-80, for example, was designed and built in 143 days—ahead of schedule. The US was at war with Germany and in desperate need of fighter-jet technology; every single day counted.

The name "Skunk Works" itself has interesting backstory. At the time, the popular newspaper comic strip *Li'l Abner* had a regular joke about a place in the forest called "Skonk Works" that brewed a potent drink made up of skunk and other foul-smelling materials. When Kelly Johnson first launched the division, there was no place for his team to work within the buildings of Lockheed. So he set up camp inside a rented circus tent next door to a manufacturing plant known for its awful stench. One of the engineers, Irv Culver, was a fan of *Li'l Abner* and answered the phone one day with "Skonk Works, inside man Culver speaking."[4] It didn't take long for the team to alter the name and start calling itself "Skunk Works," an ode of course to their rather unpleasant surroundings.

Through friends Ross heard that Lockheed Aircraft Corporation in Burbank, California, was seeking people with technical backgrounds. Her impressive credentials earned her a position as an assistant to a consulting mathematician. She helped develop fighter planes, like the P-38 Lightning, the first such plane to exceed four hundred miles per hour. While at Lockheed, she took aeronautical and mechanical-engineering courses at the University of California, receiving her California Professional Engineering certification in 1949.

In 1952 when Lockheed formed its Missile Systems Division, Ross was selected as one of its first forty employees. She was the only female engineer and only Native American in Lockheed's top-secret "Skunk Works", the beginning of Lockheed Missiles and Space Company. Ross commented: "With such a small group, you had to do everything. Aerodynamics. Structures . . . I was on

Above: Mary Golda Ross helped develop fighter planes like the P-38 Lightning, which could reach 400 miles per hour and hit 3,300 feet in a single minute, a groundbreaking and revolutionary feat at the time. The P-38 was capable of sinking a ship by carrying bombs that weighed up to two thousand pounds, and the pilots who operated these aerial weapons shot down more enemy planes than any other during WWII.

Above: The first launch of Trident I, a ballistic missile, on January 18, 1977, from Cape Canaveral, Florida.

WHAT IS THE DIFFERENCE BETWEEN A ROCKET AND A MISSILE?

The main difference between a rocket and a missile mostly comes down to tracking. Rockets have warheads and a propulsion system (like a solid rocket motor). They can travel faster and further than a traditional bomb (once released, the only thing that can impact a bomb's trajectory is gravity), but they do not have a guiding system to help them seek out a specific target. Missiles have a propulsion system and a warhead like a rocket, but they also have a guiding system. It is the guiding system that helps the missile reach its intended target via technology like radar, laser, or GPS.

Above: A Poseidon missile—one of the family of missiles that Mary Golda Ross helped to develop in the 1960s—takes off in May 1979 from a ballistic missile submarine.

the ground floor at Lockheed Missiles and Space, and I couldn't think of a more ideal situation."[5]

At Lockheed, Ross worked on defense missile systems and aided in the conceptual design of the ballistic missile systems. "Often at night there were four of us working until 11 p.m.," Ross recalled in a *San Jose Mercury News* article. "I was the pencil pusher, doing a lot of research. My state-of-the-art tools were a slide rule and a Frieden computer." Much of Ross' work during this time was classified and remains so.

When the national efforts shifted from weapons to space, Ross was there to make crucial contributions. She did research in hydrodynamics and worked on the Agena rocket, which, after its successful launch, spurred the US into the Space Age. Ross helped develop mission criteria for Mars and Venus and wrote the third volume of the *NASA Planetary Flight Handbook*, a projection of space travels for four decades. As the senior advanced systems staff engineer, she contributed to the development of the Poseidon and Trident missiles in the 1960s. She also studied flyby space probes that would ultimately examine Mars and Venus.

After retiring from Lockheed in 1973, Ross advocated for engineering and mathematics opportunities for women and Native Americans. In addition to expanding educational programs within the Council of Energy Resource Tribes and the American Indian Science and Engineering Society, she helped found the Society of Women Engineers (SWE). She cofounded the Los Angeles section of SWE and served in leadership capacities for more than a decade.

Ross has been honored by many organizations, including the Council of Energy Resource Tribes and the Silicon Valley Engineering Hall of Fame. She is immortalized in artwork at Buffalo State College in New York in the sculpture *Mary G. Ross: Scientist, Engineer,*

Cherokee-American by Lawrence Kinney and in a painting at the Smithsonian, *Ad Astra per Astra* by America Meredith, that includes references to her Cherokee heritage and to her work on the Agena spacecraft.

Ross died in 2008, just months before her one hundredth birthday, and was always grateful for everything in her life: "I have been lucky to have had so much fun. It has been an adventure all the way." Ross is buried in Park Hill, Cherokee County, Oklahoma. Her tombstone features a rocket along with the inscription "She Reached for the Stars."[2]

Above: America Meredith's painting of Mary Golda Ross entitled Ad Astra per Astra.

DOROTHY VAUGHAN

September 20, 1910 – November 10, 2008

Launching Rockets and Careers

"I guess she really didn't know she was a genius, as they said she was. To us, she was just mama."[1]

— Vaughan's daughter Ann Hammond, 2011

Above: This 1922 photograph shows members of the all-black Zeta chapter of the Alpha Kappa Alpha sorority at Wilberforce University in Ohio.

Those who worked on NASA's Scout program made a unique contribution to the US space effort. They created a launch-vehicle system that set a standard for simplicity, productivity, and reliability. They did it by establishing uncompromising standards of exactness and by an unwavering pursuit of excellence. One important member of this team was Dorothy Johnson Vaughan, a NASA mathematician who helped launch America's first satellites into space.

Vaughan was born Dorothy Johnson in Kansas City, Missouri, on September 20, 1910. When she was seven, her family moved to Morgantown, West Virginia, and by the age of fourteen, she had graduated from Beechurst High School. Dorothy received a full scholarship to Wilberforce University in Ohio, where she joined the Zeta chapter of the Alpha Kappa Alpha sorority and received a bachelor's degree in mathematics four years later, just as the Great Depression struck. One of her professors noticed her academic promise and recommended that she transfer to Howard University's graduate program, but reflecting on her parents' dire financial situation, she chose instead to teach mathematics at the segregated Robert Russa Moton High School in Farmville, Virginia. Shortly thereafter, she met and married Howard Vaughan Jr.

For the next eleven years, Dorothy taught mathematics and piano in Farmville while raising six children. In 1943 the family moved to Newport News, Virginia, where she accepted a job as a mathematician at the National Advisory Committee for Aeronautics (NACA, the predecessor agency to NASA). The Langley Memorial Aeronautical Laboratory hired her to help process data after President Roosevelt's Executive Order 8802 made nondiscrimination in defense employment official federal policy.

Above: The Robert Russa Moton High School in Farmville, Virginia, where Vaughan taught mathematics for eleven years. Students at this school played an essential role in Brown v. Board of Education, *which legally ended racial segregation in public schools in 1954.*

However, the executive order only went so far. Because Jim Crow laws that required separate work, bathroom, and eating spaces were still in practice, Vaughan was forced to work in the segregated "West Area Computing" unit with a group of black female mathematicians. As Margot Lee Shetterly writes in *Hidden Figures*, at West Computing Vaughan took her place in "one of the world's most exclusive sororities."[2] Shetterly elaborates on just how exclusive a group of that was:

In 1940, just 2 percent of all black women earned college degrees, and 60 percent of those women became teachers, mostly in public elementary and high schools. Exactly zero percent of those 1940 college graduates became engineers.[3]

Above: Vaughan sits to the far left in this 1950 photograph of a NASA social gathering.

Opposite: Vaughan and her "West Area Computers" were essential to NASA's Scout Launch Vehicle Program, which launched satellites into Earth's orbit. Shown is the first launch of Scout B in 1965.

Below: Dr. Christine Darden (shown) spent her career at NASA researching sonic booms.

Yet Vaughan not only became a mathematical engineer; in 1948, she was promoted to lead the West Computing group at Langley, making her the organization's first black supervisor. From that position she was able to collaborate with computers from other groups, make personnel recommendations, and advocate for female computers, which she did enthusiastically, regardless of their skin color.

Vaughan helmed West Computing for nearly a decade. In 1958, when NACA became NASA, segregated facilities—including the West computing office—were abolished. Dorothy Vaughan along with many of the former West Area computers joined the new Analysis and Computation Division (ACD), an integrated group of men and women on the frontier of electronic computing. Vaughan became an expert Fortran programmer, and she also contributed to the secret Scout Launch Vehicle Program, which used a rocket to launch satellites into orbit around Earth. She was able to calculate the trajectory for numerous space missions, including for the space flight of Alan Shepard, the first American in space, and the trajectory for the 1969 Apollo 11 flight to the moon.

While Vaughan sought another management position at Langley, she never received one, and in 1971 she retired from NASA. Reflecting on her career in 1994, she remarked, "I changed what I could, and what I couldn't, I endured."[4] Following her death in 2008, Vaughan's legacy of intelligence and quiet fortitude was captured by Oscar-winning actress Octavia Spencer in the film *Hidden Figures*, based on Margot Lee Shetterly's book of the same name. In addition to laying the groundwork for the successful careers of West Area "computers" like Mary Winston Jackson (see page 76), Eunice Smith, Katherine Johnson (see page 70), and Kathryn Peddrew, Vaughan's achievements at NASA paved the way for second-generation African American and female mathematicians and engineers, including Dr. Christine Darden, a leader in sonic-boom technology.

In recent years, NASA has been directed by an African American man (administrator Charles Bolden) and a woman (deputy administrator Dava Newman). Assuredly, the American space agency's current robust and diverse workforce and leadership owe a debt of gratitude to the gifted mathematician, programmer, and mother of six from Morgantown, West Virginia.

KATHERINE G. JOHNSON

b. August 26, 1918

Lives Depended on the Accuracy of Her Calculations

"Everybody was concerned about them getting there. We were concerned about them getting back."

—Katherine G. Johnson on the Apollo 13 mission

The mathematical calculations that sent Alan Shepard into space—and safely brought him home—were a matter of life and death. It was Katherine Johnson whose painstaking precision guided that 1961 mission, Freedom. Her work was essential to sending the first American into space and to America's successful space program overall.

Left: Alan Shepard returning from his 1961 space mission.

The youngest of four children, Johnson (born Katherine Coleman in 1918) was the daughter of a lumberman and a teacher and grew up in West Virginia. Signs of her mathematical talent became apparent at an early age. She told author and interviewer Margot Lee Shetterly, "I counted everything. I counted the steps to the road, the steps up to church, the number of dishes and silverware I washed . . . anything that could be counted, I did." [1,2] Johnson was so academically advanced that by age ten she was ready for high school. At fifteen she entered West Virginia State College, where she studied English, French, and mathematics.

One of her professors, W. W. Schieffelin Claytor, the third African American to earn a PhD in mathematics, told Johnson that she would make a great mathematician and offered to help her become one, even going so far as to create a course on the analytic geometry of space just for her. Johnson recognized his contribution to her success: "Many professors tell you that you'd be good at this or that, but they don't always help you with that career path. Professor Claytor made sure I was prepared to be a research mathematician." [3]

Above: Katherine Johnson in 1962 at her desk in NASA's Langley Research Center in Hampton, Virginia, complete with her adding machine and globe or "Celestial Training Device."

In 1937, at the age of eighteen, Johnson graduated summa cum laude with two bachelor's degrees: one in mathematics and one in French. She began teaching at an African American public school (one of the only options open to her at the time). She left teaching after two years when she was offered a spot at West Virginia University as part of the state's decision to integrate its graduate schools. Katherine left the graduate program early, however, to start a family with her husband, James Goble. Goble and Johnson had three daughters before he died of a brain tumor in 1956.

At a family gathering in 1952, Johnson learned that the National Advisory Committee for Aeronautics (NACA) was hiring mathematicians for its Guidance and Navigation Department. She was hired in 1953 and moved with her family to be near the Langley Memorial Aeronautical Laboratory in Hampton, Virginia. There she performed mathematical calculations, including analyzing data from flight tests, along with other women in the "pool." Her extensive knowledge of analytic geometry as well as her inquisitive nature led to a temporary assignment on an all-male flight research team. She became an invaluable member, attending editorial meetings (a taboo for women at that time) and contributing to various projects. There she performed mathematical calculations from 1953 to 1958 in the West Area computer section and the Guidance and Control Division of Langley's Flight Research Division.

Despite the fact that all those working at the division did research, Johnson and the other African American women in the computing pool had a work area—"Colored Computers"—separate from their white counterparts. Although NASA disbanded segregated work areas in 1958, disparities remained. Johnson recalled:

> *In the early days of NASA, women were not allowed to put their names on the reports—no woman in my division had had her name on a report. I was working with Ted Skopinski and he wanted to leave and go to Houston . . . but Henry Pearson, our supervisor—he was not a fan of women—kept pushing him to finish the report we were working on. Finally, Ted told him, "Katherine should finish the report, she's done most of the work anyway." So Ted left Pearson with no choice; I finished the report and my name went on it, and that was the first time that a woman in our division had her name on something.[4]*

Above: The administration building as it appeared in 1930 at Langley's Memorial Aeronautical Labratory. This is the building whete the computer pool worked.

Right: Alan Shepard—moments after splashdown—being lifted out of the water onto a U.S. Marine helicopter. Shepard's successful mission ensured his place as the first American in space.

In 1959, Johnson married Colonel James A. Johnson, a veteran of the Korean War. She contributed to the mathematics of the 1958 document *Notes on Space*, which featured lectures given by engineers in the Flight Research Division and the Pilotless Aircraft Research Division. The core of the Space Task Group was made up of engineers from these divisions, and Katherine Johnson, who had worked with many of them, joined the program as NACA became NASA in 1958. Not only did she calculate the trajectory for Alan Shepard's May 5, 1961, spaceflight, but her work helped to ensure that the Freedom 7 Mercury capsule would be quickly found after landing, using the accurate trajectory that had been established. Johnson also calculated the launch window for Shepard's 1961 Mercury mission and performed calculations for a planned mission to Mars.

So trusted was Johnson as a mathematician that when NASA used electronic computers for the first time to calculate John Glenn's orbit around Earth, Glenn requested that Johnson verify the computer's numbers by hand on her desktop mechanical calculating machine. The computers had been programmed with the orbital equations that would control the trajectory of the capsule in Glenn's Friendship 7 mission, from blastoff to splashdown, but the astronauts were leery of the electronic calculating machines, which were prone to hiccups and blackouts. Thanks in large part to Johnson, Glenn's flight was a success, and marked a pivotal moment in the Space Race between the US and the Soviet Union.

Above: The Apollo lunar landing mission profile which shows the trajectory calculated by the team at NASA, including human computer, Katherine Johnson.

Johnson also helped to calculate the trajectory for the 1969 Apollo 11 flight to the moon. In 1970, she worked on the Apollo 13 moon mission as well. When the mission was aborted after two oxygen tanks exploded, Johnson had to calculate a safe route for the astronauts still on board. She described this moment in a 2010 interview: "Everybody was concerned about them getting there. We were concerned about them getting back." Her work on Apollo 13 helped to establish a one-star observation system that would allow astronauts to determine their location in space with accuracy.

In 1986 the STEM pioneer and role model retired from Langley after thirty-three years of service. Among her many honors are the NASA Lunar Orbiter Award and three NASA Special Achievement

APOLLO 13:
"HOUSTON, WE HAVE A PROBLEM."

Intended to be the third spacecraft to land on the moon, Apollo 13 never touched down for its lunar landing. After launching on April 11, 1970, at 2:13 p.m. from the Kennedy Space Center, astronauts Jim Lovell, Jack Swigert, and Fred Haise had been on aboard for two uneventful days when a loud bang was heard on board. When the crew looked out, they saw something leaking from their spacecraft. It was, in fact, liquid oxygen. The explosion of the oxygen tank was devastating to Apollo 13, and it meant that power, water, and oxygen had to be preciously conserved or the astronauts would never make it back alive in the more than seventy-hour journey back to Earth. A moon landing was impossible with such low resources, so the goal had to change. Katherine Johnson helped calculate the trajectory Apollo 13 would use to approach the moon, but then go around it and harness the moon's gravity, so it could make its way back to Earth with low resources. At 1:07.40 EST on April 17, while the world watched with bated breath, the astronauts of Apollo 13 hit the water safely with the crew members safe, tired, and markedly thinner. NASA Deputy Administrator Thomas O. Paine said, "Although the Apollo 13 mission must be regarded as a failure, there has never been a prouder moment in the U.S. space program."[5]

Right: Apollo 13 flight directors celebrate the successful splash-down of the Command Module on April 17, 1970.

Awards. Johnson was also named Mathematician of the Year in 1997 by the National Technical Association, and Outstanding Alumnus of the Year in 1999 by West Virginia State College. In 2015, President Barack Obama said, "Black women have been a part of every great movement in American history—even if they weren't given a voice." A few months later he presented her with the Presidential Medal of Freedom—the nation's highest civilian honor—calling Johnson "a pioneer who broke the barriers of race and gender, showing generations of young people that everyone can excel in math and science."[6]

Johnson was featured in Margot Lee Shetterly's bestselling book *Hidden Figures* and portrayed in the movie of the same name by actress Taraji P. Henson. The Computational Research Facility at the Langley Research Center in Hampton Virginia bears her name, as does the Katherine G. Johnson Science Technology Institute at Alpha Academy in Fayetteville, North Carolina.

Above: President Barack Obama awards NASA mathemtician Katherine Johnson the Presidential Medal of Freedom at the White House on November 24, 2015.

When asked to name her greatest contribution to space exploration, Johnson mentions the calculations that helped synch Project Apollo's Lunar Lander with the moon-orbiting Command and Service Module. She remarked, "I liked work. I liked the stars and the stories we were telling. And it was a joy to contribute to the literature that was going to be coming out. But little did I think that it would go this far." Johnson continues to encourage her six grandchildren, eleven great-grandchildren, and other young people to pursue careers in science and technology.[7]

MARY WINSTON JACKSON

April 9, 1921 – February 11, 2005

NASA's First Black Female Engineer

"It was a segregated school, and here I am a black woman, wanting to go in there to study."[1]

—Mary Winston Jackson, 1998

The experience was all too familiar to countless ambitious, smart women of color on the civil-rights frontier. In a 1998 interview with the *Daily Press*, the seventy-six-year-old retired engineer shared some of the embarrassing aspects of making a career as an African American woman in the segregated South. Giggles erupted from her white peers after she asked, on her first day of work, where she could find the "colored" women's bathroom, which was clear across campus.

Not permitted to set foot in the whites-only cafeteria, Mary Winston Jackson was forced to order food through a window and take it back to her desk to eat. In her words, it was "sort of like when you went to the train station or the bus terminal, and you sort of put your suitcase up on the end and sat on that because you couldn't go inside. That was the cafeteria situation."[2]

Ultimately, however, Winston Jackson was not deterred from achieving what no other woman of color had achieved before. As the 1947 edition of the *Girl Scout Handbook* explained, a Girl Scouts of America member must be "ready to help out wherever she is needed. Willingness to serve is not enough; you must know how to do the job well."[3] A Girl Scout troop leader and volunteer for more than three decades, Jackson personified this tradition in her work as an outstanding mathematician, pioneering engineer, and mentor to professionals and youths.

Born on April 9, 1921, to Frank C. and Ella Scott Winston, the future engineer grew up in Hampton, Virginia, home of Hampton Institute, where civil-rights leader and Tuskegee Institute founder Booker T. Washington had pursued an education a half century earlier, following his emancipation from slavery. After graduating with the

Opposite, bottom: Jackson records aeronautical data at NASA's Langley Research Center in 1977.

Below: In 1942, Jackson graduation from the storied Hampton Institute (shown) with bachelor's degrees in mathematical and physical science.

THE HAMPTON INSTITUTE

The story of this historically black university, also known as "Home by the Sea," begins with Mary Smith Peake, a humanitarian who began providing education to the children of former slaves under an oak tree that became known as Emancipation Oak. Her first lesson as a member of the American Missionary Association (AMA) occurred on September 17, 1861. After the organization provided her with an indoor space called Brown Cottage, attendance grew to around fifty children during the day and twenty adults at night.

In 1863, the giant oak under which Peake taught former slaves to read and write became the site of the first reading of Abraham Lincoln's Emancipation Proclamation in the South, and more than one hundred and fifty years later, it still stands, an impressive ninety-eight feet in diameter, on the campus of what is now Hampton University. Officially established in 1868 on the grounds of a former plantation called "Little Scotland," it stands on the banks of Chesapeake Bay.

Four years later, an unkempt sixteen-year-old arrived on campus and eagerly swept and dusted the school's recitation room several times, hoping for an opportunity to attend the prestigious institution. This young man, who went by the name Booker Taliaferro Washington, passed the "white glove" inspection and was admitted.[4] Hampton Institute founder Samuel C. Armstrong would later help the promising student establish the famous Tuskegee Institute.

From 1878 to 1923, Hampton Institute welcomed Native American students as well, and in 1957, Rosa Parks—two years after her history-making arrest by which she earned the title "mother of the freedom movement"—began working on campus as a hostess at the Holly Tree Inn. Today, the university offers nearly one hundred degree programs, including master's programs in applied mathematics and doctoral programs in physics. It is consistently ranked among the top three historically black universities in the nation; women make up more than two-thirds of the student body.

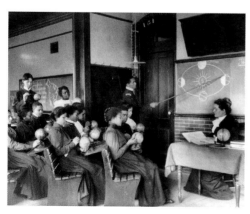

Left: This 1899 photograph shows a mathematical geography class in session at the Hampton Institute.

highest honors from the all-black George P. Phenix Training School, she continued her education at the storied Hampton Institute, earning her bachelor of science degrees in mathematics and physical science in 1942. She taught mathematics for a year at a black segregated public school in Calvert County, Maryland, and then returned to her hometown, where she became a receptionist at the King Street USO Club, an organization that served the city's black military service members during World War II.

In 1951 Winston Jackson was recruited by the National Advisory Committee for Aeronautics (NACA) as a research mathematician—or "computer"—in the segregated West Area Computing Section of the Langley Research Center. By that time she was married to Levi Jackson Sr. and had given birth to her first child, Levi Jr. She had previously met fellow mathematician and Alpha Kappa Alpha member Dorothy Vaughan (see page 66), a supervisor at Langley, while serving as army secretary at Fort Monroe.

Winston Jackson's work stood out. After two years in the computing pool, she was recruited to work for aeronautical research engineer and wind-tunnel expert Kazimierz Czarnecki. She worked with Czarnecki in the four-foot-by-four-foot (1.2 m by 1.2 m) Supersonic Pressure Tunnel, a sixty-thousand-horsepower (45,000 kW) wind tunnel used to study forces on aircraft models by producing winds at almost twice the speed of sound. Winston Jackson embraced the

Left: Jackson stands on the bottom right in this 1950s photograph of the 4 x 4 Supersonic Pressure Tunnel staff.

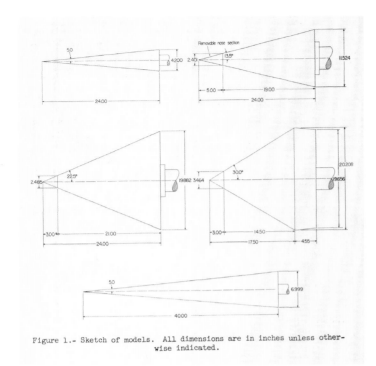

Figure 1.- Sketch of models. All dimensions are in inches unless other-wise indicated.

Above: Jackson's 1958 report on the effects of different supersonic speeds and nose angles on air flow included diagrams of the different cones they tested.

challenge of this highly specialized position, and when Czarnecki recommended she further her training to be promoted from mathematician to engineer, she jumped at the opportunity.

The University of Virginia offered graduate-level courses at night as well as in the afternoon, which made it possible for working people to attend. However, the math and physics classes required for the training were held at the then-segregated (and all-white) Hampton High School. To join her white peers, Winston Jackson had to petition the city of Hampton for special permission to attend the classes. Thankfully, her request was approved, and by 1958, she had received the promotion to aerospace engineer and become the first black female engineer in the national space agency's history. Her first report—"Effects of Nose Angle and Mach Number on Transition on Cones at Supersonic Speeds"—was also published that year.

In the 1970s, Winston Jackson helped to build a wind tunnel for the science club at Hampton's King Street Community Center so that young people could conduct experiments. In a local paper she was quoted as saying, "We have to do something like this to get them interested in science. Sometimes they are not aware of the number of black scientists, and don't even know of the career opportunities until it is too late."[5]

Much of Winston Jackson's work involved analyzing data from wind tunnel experiments and actual aircraft flight experiments at the Theoretical Aerodynamics Branch of the Subsonic-Transonic Aerodynamics Division at Langley. From 1958 until 1979, she worked as an engineer in several other NASA divisions as well—the Compressibility Research Division, Full-Scale Research Division, and High-Speed Aerodynamics Division—and authored or coauthored approximately a dozen research reports. Despite these successes, she grew increasingly frustrated at her inability to break into the male-dominated management ranks. Wanting to help other underrepresented and female colleagues, she decided to seek an administrative role outside of engineering.

In 1979, Winston Jackson received specialized training at NASA Headquarters in Washington, D.C., and then returned to Langley as its Federal Women's Program Manager, coordinating programs to help women and underrepresented minorities find and advance in careers at NASA. It was a demotion from her previous job, but it was fulfilling and meaningful work. Acting on her desire to help others, she initiated many changes at NASA, highlighted the accomplishments of underrepresented groups, and helped hire many highly qualified mathematicians, scientists, and engineers. She then helped them advance their careers, partly by advising them on how to study and how to meet the qualifications for promotion. Winston Jackson and her husband hosted many young Langley recruits working to establish their careers.

In 1969, Winston Jackson was awarded an Apollo Group Achievement Award, and in 1976 she was named Langley's Volunteer of the Year for "outstanding leadership and untiring efforts in public service and charitable organizations devoted to improving the quality of life."[6] In 1985, she retired from NASA's Langley Research Center after thirty-four years of service,

The former Girl Scout troop leader was eighty-three years old when she passed away on February 11, 2005. Not long after her death, Margot Lee Shetterly paid tribute to NASA's first black female engineer in her best-selling book *Hidden Figures*. In the film, she was portrayed by Grammy-nominated musician and actress Janelle Monáe.

WHAT IS A MACH NUMBER?

In wind-tunnel experiments, engineers determine the Mach number (M) by dividing the wind velocity at particular point on an airplane model (v) by the local speed of sound (a), which varies with temperature.

$$M = v \: / \: a$$

For example, if all the air flows over the aircraft at twice the speed of sound, the aircraft is considered to be traveling at a supersonic speed of Mach 2. Wind-tunnel experiments are necessary for measuring airflow over different parts of an aircraft, and aeronautical engineers can better understand how speed affects the aerodynamic forces—including lift and drag—on different aircraft designs, providing essential information on how to optimize air travel and improve flight safety.

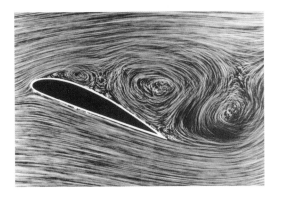

Above: This photograph of a airfoil in a wind tunnel shows the separation in air flow over the top surface.

From Circus Performer to Human Computer

SHAKUNTALA DEVI

November 4, 1929 – April 21, 2013

"Numbers have life,
they're not just symbols on paper . . ."[1]

—Shakuntala Devi

The daughter of a traveling circus performer, Shakuntala Devi often accompanied her father when he was on the road. He discovered his daughter's genius one day while they were playing cards. At the age of three she memorized an entire deck—in sequence—and beat her father in the game. By the time she was five, she was able to calculate cube roots. The young girl became the star of the show.

In Bangalore, India, where she was born, Devi and her family were destitute; her parents could not afford to give her a formal education. There were even times when

DID YOU KNOW?

Google honored Shakuntala Devi on what would have been her 84th birthday, November 4, 2013, with a "calculated" Google Doodle.

Left: Shakuntala Devi had no need for the pen and paper in her hand.

they were short of food. Instead of attending school, Devi traveled with her father, performing extreme mathematical calculations in her head, soon garnering fame, attention, and fortune for herself and her family. She began performing regularly when she was just six years old at universities in southern India. Of this time she said, "I had become the sole breadwinner of my family, and the responsibility was a huge one for a young child."[2]

Despite lacking a formal education, Devi emerged as one of the most brilliant mathematical minds of her time, and became known as "the human computer." She became an international phenomenon based on her ability to perform complicated mathematical calculations without the aid of any technological device, appearing around the world at academic institutions and theaters and on television. In 1977 at Southern Methodist University, Devi raced a

Univac computer to calculate the twenty-third root of a 201-digit number. Not only did she successfully extract the number (546,372,891) in under a minute, she beat the computer by twelve seconds.

A 1976 article in the *New York Times* reported that, in just under twenty seconds, Shakuntala Devi added the following four numbers and multiplied the result by 9,878 to get the (correct) answer 5,559,369,456,432:

25,842,278,   111,201,721,   370,247,830,   55,511,315

Professor Arthur Jensen, an educational psychologist at the University of California, Berkley, studied Devi when she visited the United States in 1988. Jensen had volunteers write problems on a chalkboard while Devi had her back turned. Once she turned around, Devi would have to solve the problem. No matter the problem, she could always solve it in under a minute. In 1990, Jensen published his studies in the journal *Intelligence*, where he wrote, "Devi solved most of the problems faster than I was able to copy them in my notebook."[4]

Devi was also a talented writer and became well known for her children's books as well as works on mathematics, astrology, and puzzles. Her books include *Figuring the Joy of Numbers*, which break down her methods, *Puzzles to Puzzle You*, which sold more than six thousand copies its first week and became popular among aspiring mathematicians, and the crime thriller *Perfect Murder*. Devi passed away in Bangalore in 2013 at the age of eighty-three.

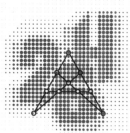

Match your wits with the
'Human Computer'
Shakuntala Devi
PUZZLES TO PUZZLE
YOU

SHATTERING RECORDS

Shakuntala Devi made it into the 1982 edition of the *Guinness Book of World Records* after multiplying two 13-digit numbers in her head—7,686,369,774,870 × 2,465,099,745,779—in just twenty-eight seconds. The numbers were picked for her at random by the Computer Department of Imperial College in London on June 18, 1980. Her correct answer?

18,947,668,177,995,426,462,773,730

Left: Shakuntala Devi, always surrounded by numbers.

ANNIE EASLEY

April 23, 1933 – June 25, 2011

NASA Rocket and Green Energy Scientist

"My mother was my greatest role model. She still is."[1]

—Annie Easley, 2001

Before laptops, desktops, or even mainframe computers, there were human computers. They were people who analyzed problems and performed calculations by hand in many areas, including the automotive, space, and nuclear-energy industries. Boundary-breaking mathematician, computer scientist, and rocket scientist Annie Easley was one of them.

Easley was born in Birmingham, Alabama, and raised by a single mother, Willie Sims, who told her that she could be anything she wanted to as long as she was prepared to work for it. Easley took this advice to heart and maneuvered, with aplomb, through Alabama's then–racially segregated schools. Despite being held after school on occasion for "talking back," she became the valedictorian of her high school. Reflecting on her childhood in a 2001 interview as part of the NASA oral "Herstory" project, she credited both her mother's steady encouragement and the influence of other high-achieving friends in Birmingham for fueling her ambition:

We weren't born with silver spoons. We came from working parents. Some of us did come from one-parent households, but we had parents who encouraged us, who raised us in a way that, this is what you need to do to keep moving ahead, and you will have to.[2]

Above: Annie Easley, 1979.

After graduating from high school, Easley moved to New Orleans to study pharmacy at Xavier University of Louisiana. She planned to move back to Birmingham after graduation, but instead she left school to get married and move to Cleveland, Ohio, where she hoped to continue her studies. As it turned out, the local university in Cleveland didn't have a pharmaceutical sciences program, so Easley decided to look for a different career. In 1955, she read a newspaper article about twin sisters who worked as human computers, providing the calculations and computations for engineers. The organization they worked for, the National Advisory Committee for Aeronautics (NACA), was seeking more people with strong skills in mathematics. Easley drove to NACA's Lewis Flight Propulsion Laboratory in Cleveland to apply for the job. Just two weeks later, she began a career that would last thirty-four years.

During the 1950s, African Americans made up less than one-fifth of a percent of the Ohio laboratory's workforce. Easley experienced racism and sexism at work, especially in the beginning when she was one of only three women of color at NACA in Cleveland. She was once cropped out of a photograph of her team displayed at an open house. But Easley, remembering her mother's advice, was undeterred.

In the "Herstory" interview she acknowledged that adjusting to her new environment was sometimes difficult, but she chose to trust in her abilities and focus on getting the job done:

My head is not in the sand. But my thing is, if I can't work with you, I will work around you. I was not about to be [so] discouraged that I'd walk away. That may be a solution for some people, but it's not mine.[3]

One of Easley's early projects was running simulations for the building of the Plum Brook Reactor Facility in Sandusky, Ohio, where staff researched nuclear-powered airplanes and, later, nuclear-powered space rockets. As part of her simulations, Easley calculated how much cement had to be poured to construct the facility. She recalls her starting salary as around $2,000 a year.

A lot changed over the years—NACA became NASA, and the Lewis Flight Propulsion Laboratory eventually became the John H. Glenn Research Center—but Easley kept up with the changing times. "I was not intentionally trying to be a pioneer," she recalls. "I wanted a job, I wanted to work." When machines began to replace the human computers in the 1970s, she pivoted to become a math technician, returning to school full-time at Cleveland State University.

In 1977 Easley graduated from Columbia State with a bachelor of science in mathematics. Getting there, though, was difficult. NASA paid the undergraduate tuition of her male colleagues—and allowed them paid leave to study—but she had to pay for the courses herself and take unpaid leaves. Easley later learned computer programming, becoming adept at writing code and implementing it in several of NASA's programs. She specialized in two computer languages: Formula Translating System (Fortran) and the Simple Object Access Protocol (SOAP).

Easley's work at NASA ended with her retirement in 1989, but her continued impact can be seen in research on energy-conversion systems and the evaluation of alternative power technology, including solar and wind technology and the battery

Below: Easley's calculations aided in the construction of the Plum Brook Nuclear Facility in Sandusky, Ohio. This 1969 aerial view shows the entire complex, including the Space Power Facility (white dome), which houses the world's largest space environment vacuum chamber.

WHAT IS FORTRAN?

Easley and other NASA mathematicians in this section made ample use of Fortran in the day-to-day work environment. Developed by IBM in 1953 and first implemented in 1957, this programming language is particularly suited to high-performance numeric computation in scientific and engineering applications. More than sixty years later, it is still used for many essential scientific purposes, including weather prediction, crash-test visualization, and crystallography (i.e., the science of determining the arrangement of atoms in crystalline solids).

technology used for early hybrid vehicles. She coauthored many papers about nuclear engines in rockets. Her contribution to the Centaur—a high-energy booster rocket that launched many communication and weather satellites and exploratory spacecraft (including Viking and Voyager)—helped shape the 1997 Cassini mission to Saturn.

Above: Fortran was originally developed for the IBM 704 mainframe computer, shown in this 1957 photograph at NASA's Langley Research Center.

In addition to her pioneering programming work, Easley volunteered as a tutor and speaker to share NASA's work and encourage STEM participation by female and underrepresented students. She and her room supervisor once made a pact to come to work wearing pantsuits in order to protest NASA's dress code for women. In 2001 she recalled that the event three decades earlier caused quite a stir, not because it violated a formal dress code, but because it was against the conventions of the time. In taking the emphasis off what a person wore to work and putting the focus on what they produced, Easley inspired other female colleagues to adopt looks at NASA that were less feminine and more utilitarian. Easley also cofounded the NASA Lewis Ski Club (she began skiing at age forty-six) and participated in other work-related activities, including the Business and Professional Women's Association, in which she remained active after her retirement until her death in 2011.

In 2001, at the age of sixty-eight, Easley spoke of her amazement at the rise of smartphone technology, her desire to learn snowboarding, and the lasting influence of her mother. She drew on the words of social activist and Harlem Renaissance poet Langston Hughes to describe her experience as a female African American in the STEM world:

> *"Life for me ain't been no crystal stair," but you got to keep struggling. You keep going because you want to.*[4]

Opposite: An illustration of the Cassini-Huygens mission to Saturn during its 1988 planning stage. With Easley's help, the spacecraft was launched nearly a decade later.

Top, left: In 1981, NASA's Science and Engineering Newsletter *featured Easley on the front cover.*

Top, right: In 1976, spurred by the energy crisis, the newly created Energy Research and Development Administration (ERDA) installed this hundred-kilowatt wind turbine for alternative energy research at Plum Brook. The hundred-foot tower supported two sixty-two foot blades that reached forty rpm in eighteen-mph winds.

MARGARET HAMILTON

b. August 17, 1936

Making Spaceflight Safe

"Looking back, we were the luckiest people in the world;
there was no choice but to be pioneers."[1]

—Margaret Hamilton, 2008

*Left: Hamilton poses with an MIT pennant
inside an Apollo lunar capsule model.*

How does one prepare for work in a field that does not yet exist? In Margaret Hamilton's case, she was well educated, fearless, and in the right place at the right time.

Before entering the technical world, Hamilton grew up in Paoli, Indiana, among family members who excelled in the humanities. Her father, Kenneth Heafield, was a poet and philosopher, and her grandfather was a Quaker minister, writer, and head schoolmaster. Their influence led her to study philosophy alongside mathematics and physics at the University of Michigan. After transferring to Earlham College—a private liberal arts school in Richmond, Indiana—Hamilton found that she was usually the only woman enrolled in her math and physics classes. However, she did have a female math professor: Florence Long. Professor Long inspired Hamilton to explore abstract math and mathematical linguistics and to consider becoming a math professor herself. There were no programming classes to take—"software engineering" was not yet a field . . . or even a term. In 1958 Hamilton earned her bachelor's degree in mathematics, with a minor in philosophy. She married lawyer James Cox Hamilton after graduating, and they had a daughter, Lauren.

In 1959, at the age of twenty-four, Hamilton moved to Massachusetts, where she intended to get an advanced degree in abstract mathematics from Brandeis University. Instead, she took a job at Massachusetts Institute of Technology (MIT) under the direction of mathematician, meteorologist, and chaos-theory pioneer Edward N. Lorenz (see page 112), who was developing software for weather prediction. With Lorenz's guidance, Hamilton taught herself hexadecimal and binary code and created her first software programs. Lorenz also encouraged her to build a platform on which her work could run. (Today we would call it a "mini operating system.") At the same time, Hamilton worked at MIT's Lincoln Laboratory, where she wrote software for the AN/FSQ-7 computerized control system (aka the XD-1), which helped the US Air Force search the skies for "unfriendly" aircraft during the Cold War.

THE SEASHORE PROGRAM

Hamilton's work on the XD-1 piqued her interest in software reliability. She recalled that when the computer crashed while executing a program, lights would flash and bells would ring; there was no hiding from it. Everybody would run into the room to find out whose program was causing the error. The computer would identify where the program failed on a very large register on its console. Hamilton recalled that her program was dubbed the "seashore program" because it sounded like waves crashing on the shore when it was running. She once got a phone call at about 4 a.m. from an operator who said that something terrible had happened to the program because it no longer sounded like a seashore. After that, she learned to debug programs using sound as a guide.

In 1963 or 1964, Hamilton was making plans to finally begin graduate studies at Brandeis University when she heard about a new contract between NASA and MIT. The university was tasked with developing the software that would "send man" to the moon, and they were assembling a team of scientists and mathematicians to write it. Hamilton immediately called the school, and within hours she had two interviews set up with the two project managers. Hamilton received a job offer from each of them. The managers flipped a coin to decide which one would get to hire her. She was thrilled that the Apollo project manager won the coin toss, and soon after Hamilton was leading the Software Engineering Division of the MIT Instrumentation Laboratory (now called the Charles Stark Draper Laboratory).

The Apollo team set about learning software engineering and systems engineering from scratch, as they helped develop the guidance and navigation system for the Apollo program. In an interview with NASA, Hamilton recounted being given total freedom and trust by upper management, who were clueless about software at the time. Like her experience in school, men outnumbered women by a significant margin, but she didn't feel she was at a tremendous disadvantage because of her gender. She recalls that the young researchers were "designing things that had to work the first time." They were making up their own rules and having a ball doing it. "The greater the challenge," Hamilton said, "the more fun we had."[2] She also made sure to involve her daughter, who came to work with her many nights and weekends.

Hamilton insisted on rigorous testing of the software programs they wrote, a commitment that proved critical to the success of the July 20, 1969 moon landing. Several minutes before Neil Armstrong's and Buzz Aldrin's Lunar Module landed on the moon's Sea of Tranquility, the crucial software overrode a manual command to switch the computer's priority system to a radar system. The software alerted everyone that the computer was shedding less-important tasks to focus on steering the descent engine and providing landing information. Had the program been different, they may have been forced to abort the approach because of computer issues. The Apollo guidance

Below: A 1969 photograph of Margaret Hamilton standing next to the navigation software that she and her MIT team produced for the Apollo Project.

software was so reliable that no bugs were discovered on any crewed Apollo mission, and it was so trusted by NASA that the code was adapted for use in other future projects, including Skylab, the Space Shuttle program, and more.

Hamilton's hard work was well rewarded, not only by the successful and safe landing on the moon by Neil Armstrong and Buzz Aldrin, but also by NASA's 2003 Exceptional Space Act Award, which recognized the value of her innovations in the Apollo software development by honoring her with the then-largest financial award ever presented to an individual. In 2016, President Barack Obama awarded her the Presidential Medal of Freedom, the highest civilian award of the United States.

Hamilton's contributions to NASA and MIT helped give birth to the field of software engineering, a term that she coined during the early Apollo missions in an attempt to elevate their discipline's status to the level of other kinds of engineering. Hamilton took all of the chances she was given and became a pioneer in a field that she helped create.

Above: Former US president Barack Obama awarded Hamilton the Presidential Medal of Freedom in 2016.

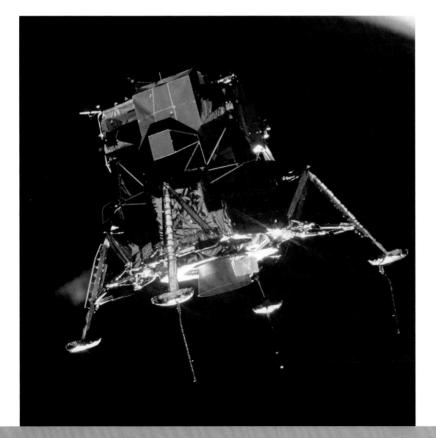

Left: The Apollo 11 lunar module with Neil Armstrong and Buzz Aldrin in a landing configuration on July 20, 1969. Without Hamilton's work on the inflight software, this historic moment may not have occurred.

PART III
Modern Math Mavens

N ow is an especially exciting time to be a female mathematician. Women have made major contributions to both pure and applied mathematics and are finally seeing their work taken seriously by the global scientific community. In 2014, a female mathematician took home the Fields Medal—math's highest honor—for the first time (see page 178). In our current social-media-dominated world, YouTube has proven to be a fertile platform for young women with a passion for mathematics, including self-described "recreational mathemusician" Vi Hart (b. 1988), who gained plenty of notoriety (and more than one million subscribers) through her lighthearted videos of math doodles and hexaflexagons. Other mathematicians have found ways to make complicated math concepts sound fun and interesting via crochet (see page 134), musical composition (see page 120), and even food (see page 172).

Opposite: Eugenia Cheng

Above: Aoibhinn Ní Shúilleabháin.

The twenty-first century has seen a dramatic shift in public perception of a field once viewed as dull and of little interest to most of the population. Shows like The *Big Bang Theory*—2016's most popular show on television for viewers eighteen to forty-nine—have glorified all things geek. Irish mathematics professor, TV presenter, and musician Aoibhinn Ní Shúilleabháin (b. 1983) has used her celebrity to combat the crusty image of scientists as "men in white coats with terrible hair."[1] American actress and mathematician Danica McKellar (b. 1975), best known for her role as Winnie Cooper on *The Wonder Years*, has also done her part to promote mathematics to young women, writing several books on the subject and taking on an animated role in the Netflix original web series *Project Mc²*.

Below: Danica McKellar at a signing for her 2008 book Math Doesn't Suck.

Women outside the media have made a difference as well, even in the face of crushing adversity. Iranian-born mathematician Sara Zahedi (b. 1981) was only ten years old when her father was killed and she was forced to emigrate to Sweden alone. As she recalls, "I didn't have any friends and I didn't know any Swedish. But math was a language I understood. In math class, I was able to communicate with my peers and I was able to make friends by solving problems with them."[2] Zahedi was the only female winner of the 2016 European Mathematical Society Prize awarded for her "outstanding research regarding the development and analysis of numerical algorithms for partial differential equations with a focus on applications to problems with dynamically changing geometry."[3]

Today several organizations—including Girls Who Code, StemBox, Blossom, Engineer Girl, and supermodel Karlie Kloss's "Kode with Klossy"—are taking square aim at bright young women with an aptitude for science and mathematics. Former first daughter Chelsea Clinton has also taken a strong stance on STEM education, traveling the country giving talks on how girls can help close the gender gap in STEM careers.

Above: Chelsea Clinton, 2016.

For girls interested specifically in math, there is the Association for Women in Mathematics (AWM), an organization founded in 1971 for the express purpose of promoting equal opportunity in the field for women and girls. Founding member and original AWM president Mary W. Gray (b. 1938) recalls one of her first actions after heading up the organization: crashing an American Mathematical Society (AMS) council meeting. After being asked to leave by members of the board, citing a "gentleman's agreement" that only board members could attend, she replied, "I'm not a gentleman. I'm staying."[4]

Five years later Gray was elected vice president of AMS, and she went on to receive honorary doctoral degrees from the University of Nebraska, Hastings College, and Mount Holyoke College after earning her law degree summa cum laude from the Washington College of Law in 1979. In 1994 the American Association for the Advancement of Science (AAAS) granted her its Mentor Award for Lifetime Achievement in recognition of "the extraordinary number of women and underrepresented minorities she has affected in her career both directly and indirectly through the influence of her former students and the programs she has initiated and developed."[5] In 2001 she received the Presidential Award for Excellence in Science, Engineering and Mathematics Mentoring, and in July 2017 she was given

Above: Rhonda Hughes (page 105), Ingrid Daubechies (page 126), and Sylvia Bozeman (page 102) at the Association for Women in Math Awards in 2015.

the Karl E. Peace Award for Outstanding Statistical Contributions for the Betterment of Society.

In its nearly fifty-year existence, AWM has sponsored numerous programs, workshops, and prizes in furtherance of its mission, including three honorary lecture series. The Noether Lectures, named in honor of abstract algebra founder Emmy Noether (see page 36), have been given annually since 1980 in honor of women who have made "fundamental and sustained contributions to the mathematical sciences." The Falconer Lecture series, begun in 1996, commemorates African American mathematician and educator Etta Z. Falconer—specifically, her "profound vision and accomplishments in enhancing the movement of minorities and women into scientific careers." Finally, celebrating the nineteenth-century Russian mathematician Sofia Kovalevskaya (see page 26), the Kovalevsky Lectures have been given annually since 2003 in honor of women who have "made distinguished contributions in applied or computational mathematics."[6]

Many of the women featured in this section have a connection of some kind to AWM, whether they have given honorary lectures or joined the executive committee or advisory board. Like the mathematical pioneers of Part I, they also represent incredible mathematical potential from many corners of the world, from Belgium to Iran to the American South.

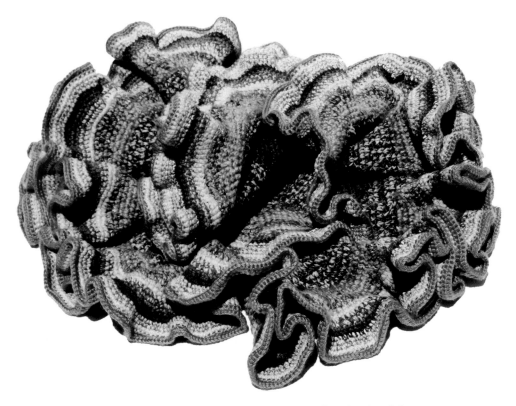

Above: Daina Taimina (see page 134) uses crochet models like these to teach students about hyperbolic geometry.

SYLVIA T. BOZEMAN

b. 1947

*Appointed to the President's Committee on the
National Medal of Science by President Barack Obama*

"I've always loved math, but I'm equally passionate about
supporting and encouraging women to get advanced
degrees in math—especially women of color."[1]

—Sylvia T. Bozeman

Sylvia Bozeman (born Sylvia Trimble) began her brilliant math career in a one-room schoolhouse in Camp Hill, Alabama, a rural area with segregated schools. Little did she know back then that she would one day be appointed to the President's Committee on the National Medal of Science. If you ask her, though, Bozeman will tell you she believes that there is no limit to what women can achieve if they are encouraged and supported. After all, she is the perfect example.

While attending the segregated primary and secondary schools in rural Alabama, Bozeman was encouraged by her teachers and parents, who, she says, "instilled in me a love of learning and a concern for the education of others."[2] Her father was an insurance agent and her mother the primary caregiver to five children. "My mother only finished the 12th grade," Bozeman remembers, "but she was always excited about math."[3] That excitement spread to her daughter, and when Bozeman was in high school, a math teacher recognized the talent of his students—including Bozeman—and offered a trigonometry class after school to better prepare them for college level courses.

Above: An early shot of Bozeman's parents (now deceased), Horace Trimble, Sr. and Robbie Trimble.

The valedictorian of her high school class, Bozeman entered Alabama A&M University to study mathematics in 1964. Howard Foster, then chair of the department of mathematics and physics, employed her as his research assistant and introduced her to the computer at NASA (Marshall Space Flight Center). He also made her a teaching assistant at his summer science program for high school students. Bozeman gained additional mathematics and computing experiences through a special summer program at Harvard University following her junior year. It was at college that she met fellow math major Robert Bozeman. They were married in 1968, shortly after graduating from college—Sylvia as salutatorian—and both went on to make history.

Below: The history-making couple, Robert and Sylvia Bozeman, hope to celebrate fifty years of marriage in 2018.

Just one year after the graduate mathematics program was integrated at Vanderbilt University, the couple began graduate studies there. Bozeman went on to become the first African American woman to earn a master's degree in mathematics from the university in 1970. The couple had a son and daughter while Sylvia Bozeman taught part-time at Vanderbilt and Tennessee State University. Meanwhile her husband finished his doctorate, the first African American to earn a PhD in mathematics at Vanderbilt.

The couple found teaching positions in Atlanta, Georgia. Sylvia Bozeman began teaching at Spelman College (a historically black women's college) in 1972. After two years, Bozeman decided that she needed to obtain a PhD in order to advance her career. Fewer than twenty black women in America held doctoral degrees in mathematics at that time. In the late 1970s, she took a three-year leave of absence from Spelman to attend Emory University

Above: Sylvia Bozeman, now Emerita Faculty, spent more than thirty-five years at Spelman College, including ten years as chair of the mathematics department. Photo courtesy of Vanderbilt University.

Above: Spelman College sign detailing the history of the women's college.

in Atlanta, where she earned her doctorate with a dissertation entitled "Representations of Generalized Inverses of Fredholm Operators."

Bozeman remained at Spelman for more than thirty-five years as a professor, where other roles at various times included director of the Center for the Scientific Applications of Mathematics, and associate provost for Science and Mathematics. She also served as an adjunct faculty member in mathematics at Atlanta University from 1983 to 1985. Her noted scholarship includes publications and research funded by organizations such as NASA and the US Office of Army Research.

In 1998, Bozeman collaborated with Rhonda Hughes to create a transition and mentoring program for women entering graduate school in the mathematical sciences. The Enhancing Diversity in Graduate Education program, or EDGE, has made a significant impact on the mathematics and scientific communities and was given special recognition by the American Mathematics Society for its effectiveness. "I am very proud that this idea has been sustained because of the impact of the EDGE Program on the success of women participants and the resulting enhancement of professional collaborations among women," Bozeman said.[5]

Bozeman is a member of the Mathematical Association of America (MAA), the country's largest organization of its kind for college and university professors, and in 1997, she became the first African American elected as an MAA Southeastern Section Governor. That same year, she was selected to shepherd the most expensive

GIVING WOMEN AN EDGE

Bozeman believes that one of the major factors contributing to success in graduate programs—and in life—is support: being part of a team or community can improve resilience and persistence. The lack of diversity in many of the STEM fields, including math, often results in isolation and loneliness for graduate students from underrepresented groups. The Enhancing Diversity in Graduate Education program was launched by Sylvia Bozeman and Rhonda Hughes in 1998 to help female students in graduate programs get the support they need to finish their advanced degrees. Mentoring, support, and preparation are all essential components to increasing and retaining women in STEM programs and jobs. Since EDGE began, at least eighty of the more than 200 EDGE participants have already earned doctoral degrees and entered a variety of careers, with many more still in the pipeline. The EDGE program was honored in 2007 by the American Mathematics Society for its effective results.

Above: Sylvia Bozeman and Rhonda Hughes (seated) with a group of EDGE (Enhancing Diversity in Graduate Education) students.

construction project in the history of Spelman: the $25 million Spelman College Science Center. In 2016, President Barack Obama appointed Bozeman to the President's Committee on the National Medal of Science. The twelve-member committee is responsible for selecting award nominees for the president's consideration. Bozeman said, "I want to cultivate the next generation of female leaders and innovators in mathematics—and I want to help the breakthrough work of these women to gain recognition with honors, including the National Medal of Science."[6]

Among Bozeman's many accolades are being named a fellow of the American Association for the Advancement of Science in 2010 and a fellow of the American Mathematical Society in 2013. In 2012, she received a Lifetime Service Award from the National Association of Mathematicians.

Ever grateful for those who helped her along the way, from her teachers to her family, Bozeman told an interviewer, "I am most proud that with the help of family and community, and the grace of God, our two children now have their own families and careers."[7] In addition to her career, Bozeman is also committed to her family and active in her church, where she has played in a handbell choir for more than thirty years.

Bozeman is committed to remaining connected with mathematics through teaching and scholarship while at the same time promoting the development of women in mathematics. She advises young people to find proper mentors in both professional and personal realms: "Listen to the advice of others, filter it and use what you can."[9]

Opposite: Bozeman's granddaughter makes her a rocket to use in a HistoryMakers elementary school visit.

FERN Y. HUNT

b. January 14, 1948

Connecting Mathematics and Technology

"The applications of mathematics can be totally different, yet the mathematics itself can be the same and very interesting in its own right—such is its universal quality."[1]

—Fern Hunt, 2017

Many sparks ignited Fern Hunt's interest in science and mathematics. There was the chemistry set she received for Christmas at age nine and the frequent forays from the Amsterdam Housing Projects to visit museums and libraries around New York City. Ninth grade at La Salle Junior High brought Freda Denenmark's algebra class and an encouraging science-club advisor, Charles Wilson, who suggested Hunt attend the Saturday Science Program at Columbia University. Hunt took his advice, taking a course in mathematics. Wilson also urged her to apply to the highly selective Bronx High School of Science, which she later attended.

Left: A woman on the New York City subway in 1973. When Hunt was growing up on Manhattan's upper west side, the city had high rates of poverty and crime.

Although she was never a fan of tests or "writing down just the answer you need to get a good grade," Hunt says she "always enjoyed learning and discussing ideas with people."[2] By the time she was fifteen, she knew that she would pursue mathematics.

Hunt had support at home as well. Before World War I, her grandparents had immigrated from Jamaica in search of opportunity in the Big Apple. Neither her father, a mail handler, nor her mother, a transcribing typist, were inclined toward math or science, but they supported Hunt's interests. Her mother, having attended Hunter College for two years until it became too costly to continue, also encouraged her to pursue higher education. She did, earning a bachelor's degree in mathematics from Bryn Mawr College in 1969. During her junior year, her professor Martin Avery Snyder urged her to attend the graduate school he had attended, the Courant Institute of Mathematical Sciences at New York University. Hunt followed his advice, and by 1978 she had completed a PhD in mathematics.

Hunt worked at the University of Utah before moving back east to be closer to her family. She spent fifteen years in the mathematics department of Howard University and then accepted a position at the National Institutes of Health in the Laboratory of Mathematical Biology. She now works at the National Institute of Standards and Technology in the computing and applied-mathematics laboratory, studying the physical, chemical, and electrical properties of materials important to US industry.

Below: Students Aspire, *a public sculpture by Elizabeth Catlett on the campus of Howard University, where Hunt taught mathematics for fifteen years.*

THE MONTE CARLO METHOD

One of Hunt's research areas is probability theory, which she uses to help scientists analyze real-world problems, like the structure of disease bacteria and how to use computer algorithms and light physics to visualize the appearance of objects that don't exist. She has also applied mathematics to the question of how the genetic makeup of species changes in the presence of a deteriorating environment, such as when animals overgraze.

One of the primary techniques she uses in her probability research involves the Monte Carlo method. In extremely simple terms, this approach uses random sampling computer algorithms to calculate risk and simulate complex systems. While there are many different variations on this method, they tend to follow four basic steps:

1. Define the input domain
2. Distribute inputs randomly over the domain
3. Calculate an algorithm based on their distribution
4. Use this algorithm to calculate your prediction or model

In reality, coming up with an accurate prediction via the Monte Carlo method for anything worthy of study is not very practical without the aid of a computer, which can use special algorithms to place thousands of random points and provide estimates close to a few hundredths of a percent of the actual value. And once you have come up with a good mathematical formula for predicting data points based on a set of variables, you can begin to build models for everything from microscopic biomolecules to complex weather patterns.

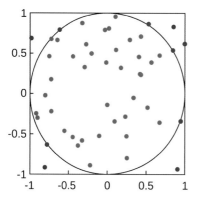

Left: Forty of the fifty randomly scattered dots fall within the inscribed circle domain, giving us a probability ratio of 0.8, which is close in value to the ratio of the circle domain's area to the square domain's area ($\pi/4$ or 0.785). When the area of a domain is not known, this random scatter method can help determine it.

Left: An aerial veiw of Bethesda, Maryland, headquarters of the National Institutes of Health, 2003.

In 2000, Hunt was honored with the prestigious Arthur S. Flemming Award for Outstanding Federal Service for her contributions to mathematical biology, computational geometry, nonlinear dynamics, and more. She was applauded for the impact of her extensive close collaborations with scientists and engineers seeking to apply these developments to diverse scientific and technological problems. Examples include flow in complex geometries, modeling of micromagnetic devices, study of optical reflection, image rendering in computer graphics, and visualization of genetic sequences. Hunt was also cited for her outstanding dedication to the mathematics profession. She is involved with several professional mathematical societies and has served as a consultant for organizations like the US Department of Energy. An active member of scientific and local communities, Hunt has been a mentor and leading proponent of careers in mathematics for students at the high school, undergraduate, and graduate levels, especially for women and minorities.

Hunt is a frequent presenter at meetings of the American Mathematical Society and the Mathematical Association of America. "I have tried to encourage and inspire young people to enjoy mathematics and to understand its power when I was teaching and in participation in programs that encourage and advise aspiring mathematics students, particularly those of color,"[3] says Hunt, who also participates in the Infinite Possibilities Conference and the HistoryMakers project. She is also actively involved in the Conference of African American Researchers in the Mathematical Sciences,

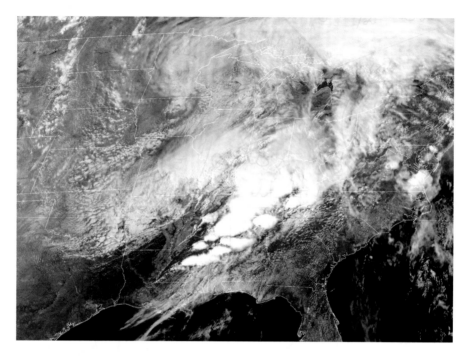

Above: A satellite image of the weather system that spawned the deadliest tornado outbreak since 1974. On April 27-28, 2013, more than 250 people were killed across six states.

CHAOS THEORY

One of Fern Hunt's major interests is chaos theory, which involves the mathematical study of complex dynamical systems like the stock market, weather, migratory patterns, or the spread of vegetation across a continent. These patterns involve so many interrelated variables that they cannot be easily predicted without giant, high-powered computers analyzing mountains of data. The term *butterfly effect*—coined by mathematician Edward Lorenz to describe how the flapping wings of a butterfly in, say, Peru could theoretically affect the course of a tornado in Oklahoma several weeks later—is often evoked when talking about chaos theory. Because of the apparent randomness of dynamical systems like weather that impact our daily lives, a lot of research has gone into finding better methods of prediction. Alas, despite enormous leaps in computer processing power and the ever-increasing sophistication of mathematical models, we are in many ways still at the whim of Mother Nature.

a board of trustee member at her alma mater (Bryn Mawr), and a teacher in a summer program designed to encourage women and minority students to pursue graduate study in mathematics.

The advice she would give her younger self is helpful for anyone contemplating their future, in any field:

> *Be much braver, put yourself forward and get beyond your comfort zone. Hang out with the people who are doing the kind of mathematics you want to do—even if they aren't that much into you. Make use of conferences, websites, journals, meetings, whatever method that helps you create ideas and correct your mistakes the fastest.*
>
> *To do this, of course, you need a core—a definition of who you are—that stands, regardless of the opinion of others, including those you admire.*[4]

Hunt plays the piano and enjoys the theater. She finds that reading history has helped to give her story—its successes and heartbreaks—a context. "The tragedies of American and European history that haunt and limit me are easier to bear," she says. "Reading about the faith and determination of my forebears strengthens me to struggle, to reach out to others as we struggle as we do and will overcome them."[5]

Above: Hunt also has a talent for music. Here, she plays the piano and sings with Sunday School students.

MARIA KLAWE

b. 1951

Mathematician, Computer Scientist,
Harvey Mudd College President, STEM Advocate

"Hard work and persistence is more important for success in mathematics than natural ability. Actually, I would give this advice to anyone working in any field, but it's especially important in areas like mathematics and physics where the traditional view was that natural ability was the primary factor in success."[1]

—Maria Klawe

One day, Maria Klawe and a friend were sitting on a cliff overlooking the city of Vancouver, British Columbia. As they admired the cityscape, the mountains, and the ocean, her friend asked, "Don't you wish you had been born fifty years earlier so you could have lived here before Vancouver became a big city?"

Klawe responded with an emphatic "No." If she had been born fifty years earlier, she explained, her life would have been miserable. As a woman, it would have been almost impossible to have a career in mathematics, technology, or computer science, or to become a powerful figure in academia. And, she said, she certainly couldn't have combined these careers with marriage and two children.[2]

Since 2008, Klawe has served as the first female president of Harvey Mudd College, one of the premier engineering, science, and mathematics colleges in the United States. Before coming to Harvey Mudd, she served as the first female dean of engineering at Princeton University and the first female in several leadership positions at the University of British Columbia. Klawe also spent eight years in industry, serving at IBM Almaden Research Center in San Jose, California, first as a research scientist, then as manager of the Discrete Mathematics Group and manager of the Mathematics and Related Computer Science Department.

As the second of four daughters, Klawe says she was always viewed by everyone in the family as "the boy." She recalls, "For a long time (probably 'til around age nine) I believed I would wake up one morning as an actual boy." She says her father nurtured and supported her interests in "boy-related" activities, and both parents believed she could excel in any academic area or leadership role.

When Klawe was in high school, her teachers routinely indicated that "girls couldn't do math or physics." This never rang true for Klawe, however. "Everything we were being taught in math seemed natural and incredibly straightforward," she said. In college, in a course that taught epsilon-delta proofs for calculus, she fell in love with the beauty of how everything fit together and how calculus could be used to solve so many problems. Some professors would ask her why she wanted to be a mathematician since, according to them, there were so few women role models in the field, while other educators found it refreshing to have a female student who loved math. Their encouragement led her to get involved with K–12 outreach activities to convince girls, teachers, and parents that girls can excel in math and science and that doing well in high school math is important for success in any professional career. Klawe received her PhD (1977) and BS (1973) in mathematics from the University of Alberta.

Above: President Maria Klawe, standing outside of her office at Harvey Mudd College.

WHAT IS IMPOSTER SYNDROME?

First coined by clinical psychologist Pauline Clance and colleague Suzanne Imes in 1978, Imposter Syndrome—also known as Fraud Syndrome—is an internalized belief or notion that you are not successful or capable of high achievement, that you might not deserve any success you may have, and that you'll be exposed as a fraud or just lucky. It's a condition that affects many successful women, including Maria Klawe, who has spoken extensively on the subject, as well as Emma Watson, Kate Winslet, Sheryl Sandberg, Jennifer Lopez, Lady Gaga, Tina Fey, Amy Poehler, and many others.

As a teacher, Klawe's focus was on making mathematics accessible and appealing to all students and using technology to enhance learning and motivation. In 2005, she won the Princeton Engineering Student Council teaching award for her work teaching calculus. She also organized the Aphasia Project at the University of British Columbia, bringing together faculty from human-computer interaction, psychology and audiology, and speech sciences to produce handheld devices to improve the quality of life and independence of people with aphasia (loss of language most commonly caused by stroke).

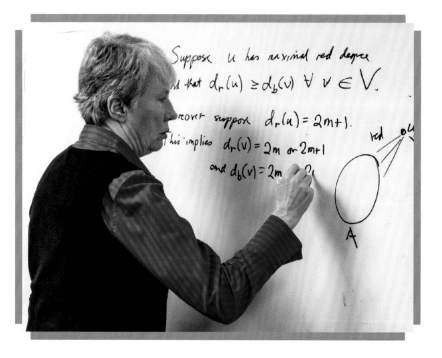

Right: Maria Klawe felt that the mathematics curriculum wasn't accessible to everyone, so she set out to change that.

Klawe has been active in many organizations promoting women and leadership in science and technology. She has chaired the Anita Borg Institute of Women and Technology, was the founding co-chair of the Computing Research Association's committee on the status of women, and is a fellow of the American Academy of Arts and Sciences, the American Mathematical Society (past trustee), and the Association for Computing Machinery (past president). She's lent her expertise to many boards, including nonprofit Math for America.

The recipient of the 2014 Women of Vision ABIE Award for Leadership, Klawe was ranked seventeen on Fortune's 2014 list of the World's 50 Greatest Leaders. In 2015, she was honored with both the Lifetime Achievement Award from the Canadian Association of Computer Science and the Achievement Award from the American Association of University Women, and she was inducted into the US News STEM Solutions Leadership Hall of Fame. The following year Klawe was honored with the Computing Research Association's 2016 Distinguished Service Award.

As a senior woman in technology, Klawe is inspired by the opportunity to help create positive change and gives talks at

Above: Besides loving math, Klawe also loves painting (see page 119) and skateboarding.

KICKSTARTING
FEMALE PARTICIPATION IN STEM

Maria Klawe has been a longtime advocate for women in STEM. While she was the dean of science at the University of British Columbia, the number of women faculty members doubled from twenty-four to forty-eight. During the five years she chaired the NSERC-IBM for women and engineering, Klawe saw an increase in female computer-science majors from 16 to 27 percent and the number of faculty in the computer-science department from two women to seven. When Klawe arrived as president at Harvey Mudd College in July 2006, females composed about 30 percent of faculty and staff. By 2012, 45 percent of the students and 40 percent of the faculty were female. And as Klawe herself says, "As the first female in so many of my positions, I have faced more suspicion and scrutiny than many male leaders would . . ."[3]

Above: Maria Klawe believes that mathematics will be needed to solve all the major problems happening in the world right now.

international conferences, national symposia, and colleges across the US and Canada about diversity in science, technology, engineering, and mathematics as well as gender and gaming, weaving in lessons from her own career in the STEM industry and education. She believes the opportunities in these areas far outweigh the difficulties and shares this message around the world. "Hard work and persistence is more important for success in mathematics than natural ability. Actually, I would give this advice to anyone working in any field, but it's especially important in areas like mathematics and physics where the traditional view was that natural ability was the primary factor in success."

In recent years Klawe has devoted particular attention to improving K-12 science and mathematics education. To encourage young people in math, she shows them problems that are completely different from the kinds they are learning in school but can be solved applying the knowledge they have. She also shows them how mathematics is used in computer animation. Their involvement in the discipline, she believes, is essential for the future. "Progress in addressing almost every major problem facing the world (e.g., climate change, healthcare, education) will involve mathematics in some form, though many other disciplines will be involved as well. Mathematics also plays a major role in machine learning and data science, and these disciplines are revolutionizing every aspect of society."

"Ask for help and take it . . . surround yourself with people who encourage you, share your feelings with others, celebrate your successes, be willing to try new approaches if your usual one isn't working, and don't let your fears stop you from giving your best effort. . ."

—Maria Klawe

In her talks, Klawe often mentions her painting practice, something she has enjoyed since her youth and developed during college. In the past, she kept her painting hobby secret because she worried it would undermine her credibility as a female mathematician and computer scientist. She now does art "on the side" but no longer feels the need to hide it. Her paintings adorn her office, and she's even shown them at a gallery and to students she's recruiting. "I want them to know that many leading engineers and scientists are artists, musicians, dancers or writers. Engineering and science are creative disciplines. It shouldn't be surprising that the creative energy, passion and talent cross into other areas."[4]

Starling Flox painted by Maria Klawe in 2005.

AMI RADUNSKAYA

Leader, Composer, Advocate

"Are music, mathematics, growth and evolution articulations of the same laws?"[1]

—Ami Radunskaya, 2015

Ami Radunskaya says she's always loved patterns. An accomplished cellist, she often makes music from them. Growing up in California, she used to find hours of enjoyment pressing the buttons and watching the wheels go around on an adding machine in her father's office. When she saw patterns emerging from a mathematical formula on a dot-matrix printer, she thought it was "magic" and found herself wanting to learn more about fractals, dynamics, and chaos. But the study of mathematics took a back seat to her musical aspirations.

She began playing the cello at the age of nine; when she graduated from high school at age sixteen, she worked as both a cellist and a music composer, spending seven years as a member of the Oakland Symphony. Radunskaya also performed throughout the United States and Europe with Don Buchla, a pioneer of electronic music.

After ten years in music, Radunskaya, a single mother, began her undergraduate studies at the University of California, Berkeley. There, she studied computer science and chemistry before eventually majoring in mathematics. In 1992, under the supervision of Donald Samuel Ornstein, she earned her PhD in mathematics from Stanford University with a dissertation entitled "Statistical Properties of Deterministic Bernoulli Flows." When Radunskaya left California for Texas to pursue her postdoctoral studies at Rice University, she was the only woman in the school's mathematics department. Three years later, she returned to California to join the mathematics faculty of Pomona College, specializing in ergodic theory, a discipline of mathematics that focuses on dynamical systems and their applications to various real-world problems.

For the last fifteen years, Radunskaya has been applying dynamical systems to various medical problems. Her research projects involve mathematical models of cancer immunotherapy, the effect of anti-coagulants, and stochastic phenomena. She spent a recent sabbatical

Opposite, below: A view of UC Berkeley's campus next to San Francisco Bay, where Radunskaya majored in mathematics.

Left: The Hoover Tower at Stanford University, where Radunskaya got her mathematics PhD.

WHAT ARE DYNAMICAL SYSTEMS?

Dynamical systems are mathematical models for evolving physical phenomena. These models are used for three main purposes: to predict, to diagnose, and to understand. In predictive or "generative" models, the aim is to predict the future from observations of the past and present. A common application is economic forecasting. In diagnostic models, the aim is to figure out what in the past might have led to the present state—in essence, reasoning backward from effects to causes. The most obvious application is medical diagnosis, in which the effects are the observed symptoms and the causes are diseases. In the third category, the model offers insight into how something works, as in the case of a scientist offering a theory for a particular chemical reaction in terms of a set of differential equations involving temperature, pressure, and various compounds.

In the case of highly stochastic (i.e., unpredictable) phenomena, like the decay of certain radioactive material, coming up with a mathematical model is particularly challenging. Other phenomena, such as stock-market or weather patterns, require equations that are so complex or so critically dependent on extremely precise measurements that accurate long-term forecasting is near impossible.

in a school of pharmacy studying strategies for delivering drugs to the brain. She is also investigating the influence of social networks on friendships, as well as the spread of ideas and disease. Radunskaya's musical background continues to influence her work, too. She remains passionate about the intersection of mathematics and music, especially the ways in which dynamical systems can be applied to instrument modeling, sound generation, and interactive composition.

As president of the Association for Women in Mathematics (AWM), Radunskaya hopes to encourage young women to turn their enjoyment of puzzles and numbers into a career, just as she has. "Our community will benefit from the recruitment and retention of talented young women who dream of doing mathematics for a living,"[2] she says. Her strong belief in the power of collaboration has inspired her to organize sessions at national mathematics meetings that celebrate diversity. She has also pushed to include more mathematicians from underrepresented groups as symposium speakers. In 2013, working together with University of Michigan professor Trachette Jackson (see page 160), Radunskaya organized a research workshop at the University of Minnesota's Institute of Mathematics and its Applications called WhAM!: Women in Applied Mathematics. Additionally, she produced four Women in Math in Southern California (WiMSoCAL) research symposia with the purpose of encouraging collaboration between graduate students, early-career mathematicians, and senior women.

Above: A close-up of Don Buchla's 200e Electric Music Box model. Buchla created his ca. 1978 Sili-Con Cello model for Radunskaya, who performed with him in the US and Europe.

Radunskaya's work as a mentor to up-and-coming mathematicians is well recognized. In 2016, she was awarded the American Association for the Advancement of Science (AAAS) Mentor Award for launching "dramatic education and research changes leading to an increase in the number of female doctorates in the field of mathematics."[3] In an article published at the time, the association commended her work supporting women—especially women of color—pursuing PhDs in mathematics. In her time as a mentor leading up to the award in 2016, AAAS noted, Radunskaya had helped eighty-two mentees earn mathematics PhDs. Of those, eighty were female and two were male. Twenty-three of them were African American and five were Latino.

In addition to her work with the Association for Women in Mathematics and the Institute of Mathematics and Its Applications, Radunskaya is a codirector of the EDGE (Enhancing Diversity in Graduate Education) program, which helps women successfully complete PhD programs in mathematics and places them in visible leadership positions throughout the mathematics community. Her impressive list of professional honors also includes a 2004 Irvine Fellowship for Excellence in Faculty Mentoring and a 2012 Pomona College Wig Award for Excellence in Teaching.

The 2014 documentary *The Empowerment Project: Ordinary Women Doing Extraordinary Things* featured a thirty-day, seven-thousand mile (11,265 km) cross-country road trip to interview eight incredible women, including Radunskaya as well as four-star navy admiral Michelle Howard and astronaut Dr. Sandy Magnus. The purpose of the documentary was to challenge girls to ask themselves, "What would I do if I knew I would succeed?" by showcasing ambitious women, like Radunskaya, who excel in male-dominated fields.

Opposite: Inside the Mabel Shaw Bridges Hall of Music at Pomona College, where Radunskaya is professor of mathematics.

INGRID DAUBECHIES

b. August 17, 1954

Fixing Art and Medicine with Mathematics

"If I have always wanted to learn more it was because different types of problems require different types of mathematical tools. I use and develop math to approach problems."[1]

—Ingrid Daubechies, 2014

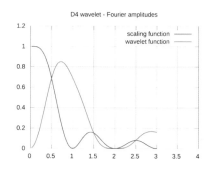

N ine-year-old Ingrid Daubechies couldn't get to sleep, so she played her favorite game: computing the powers of 2 in her head: 1, 2, 4, 8, 16. . . . Fascinated by how quickly they grew, she multiplied each number by 2 as far as she could before dozing off.

This early exercise in exponential growth, along with an interest in machinery and crafting, enriched her childhood. Daubechies loved sewing clothes for her dolls, turning flat fabric into curved pieces. Her father, Marcel, was a civil mining engineer, and her mother, Simone, a criminologist and historian. Together they encouraged their daughter to pursue a career in the sciences.

Born in Houthalen, Belgium, Daubechies completed her early education and attended Vrije Universiteit Brussel at age seventeen. She earned her bachelor's degree in physics and made frequent visits to the CNRS Center for Theoretical Physics in Marseille, France, while she worked on her dissertation "Representation of Quantum Mechanical Operators by Kernels on Hilbert Spaces of Analytic Functions." In 1980, Daubechies earned her PhD in theoretical physics.

Opposite, below: This graph shows the amplitudes of the frequency spectra for Daubechies' 4-tap wavelet.

WHAT IS QUANTUM MECHANICS?

In 1900, physicist Max Planck put forward his revolutionary quantum hypothesis that energy is radiated and absorbed in discrete packets he called "quanta." Five years later Albert Einstein used Planck's theory to explain the *photoelectric effect* (i.e., the emission of electrons when light shines on a material). Since that time, important developments have been made by the likes of Erwin Schrödinger (of "Schrödinger's cat"* fame) and Werner Heisenberg (who developed the Heisenberg uncertainty principle[2]). Today, the field of quantum mechanics, with its insights on the nature of atoms and subatomic particles, has made possible the development of many important modern technologies, including medical and research imaging, lasers, and computer microprocessors.

---

*The famous Schrödinger's cat thought experiment was devised by the Austrian physicist in 1935. In it he asks subjects to imagine a cat concealed in a box with a flask of poison, a Geiger counter, and a radioactive substance. If the Geiger counter detects even one radioactive atom decaying, the flask will be shattered, releasing the poison and killing the cat. According to one interpretation of quantum mechanics, the cat is considered both dead and alive (i.e., in a state of quantum superposition) because there's no knowing when the random decay event will occur, and the external observer has no way of knowing whether it has or has not occurred. With this scenario, Schrödinger challenges us to ask the question, "When does quantum superposition end and reality begin? When I peek in the box and see either a dead cat or a live cat?"

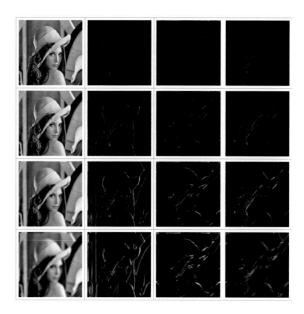

Above, left: An elliptical building on the Ixelles campus of Vrije Universiteit Brussel, where Daubechies majored in physics.

Above, right: Example of a wavelet tranform applied to an image of varying sharpness. Daubechies' wavelets make it possible to compress image sizes without losing too much detail.

In the early 1980s, French mathematician Jean Morlet made an important discovery related to analyzing geophysical signals. Rather than using Fourier transform methods to break signals into their constituent frequencies, he and fellow scientist Alex Grossmann developed wavelet transforms, which provide information about when these frequencies occur and for how long. In 1985, Daubechies worked with Grossmann and Yves Meyer to enable wavelet functions to be reconstructed from a discrete set of values, and a year later, while a guest researcher at the Courant Institute of Mathematical Sciences in New York City, she made an important discovery when she managed to construct compactly supported continuous wavelets. This invention allowed wavelet theory to make significant impacts on digital signal processing and image compression.

Daubechies has also had an enormous impact the art world, collaborating with many artists, historians, and conservators to restore ancient works of art and identify forgeries. She and her team use her wavelet transforms to perform virtual restorations of high-resolution digital versions of paintings by mathematically detecting and removing cracks in the paint layer caused by age. Paleographers can then decipher more of the painting, leading to important discoveries in various works of art, including the fourteenth-century *Ghent Altarpiece* and disputed paintings of Vincent van Gogh. In a 2014 interview with The World Academy of Sciences, Daubechies explained the process in detail:

WHAT CAN
WAVELET ANALYSIS DO?

Innovations related to wavelet analysis have a number of applications. For example, the FBI used wavelet compression to store and retrieve its 250 million fingerprint files that required 250 terabytes of space. A typical fingerprint file could occupy about 10 megabytes of space, but wavelet compression allowed the FBI to reduce the amount of computer memory needed for fingerprint records by 93 percent. As Daubechies explained in 2014,

> *Wavelets decompose the image into building blocks of different scale that, together, describe what's going on in the image. Simplifying a lot, this approach tells you where you need to put lots of detail (because pixels differ a lot from their neighbors) and where not, in image analysis.*[3]

In the medical field, scanner-based imaging systems, such as magnetic resonance imaging (MRI), also benefit from wavelet technology, which can improve poorly digitized images of a body's internal structures. As a result, patients are exposed to less radiation during scanning, making the imaging process quicker, less expensive, and safer.

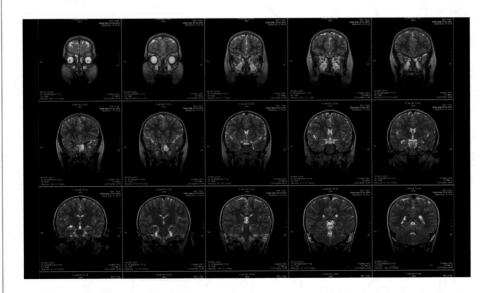

In the case of the **Ghent Altarpiece** *we combined three detection methods to make a map of all the cracks. Then we virtually in-painted these cracks, thus reconstructing a sharper view of, in this particular case, the letters in a medieval book depicted in the background on one of the panels. This made it possible for paleographers to decipher many more words than the two they could read before; as a result they can now unambiguously identify . . . that the painter referred to a particular text by Thomas of Aquinas about the Annunciation.*[4]

Below: The restored Ghent Altarpiece (ca. 1431) by Flemish brothers Hubert and Jan van Eyck.

Daubechies settled in the US permanently after marrying fellow mathematician A. Robert Calderbank in 1987. That same year she became a technical staff member at the mathematics research Center of AT&T Bell Laboratories in New Jersey. She then returned to academia, teaching at both the University of Michigan and Rutgers University. In 1992, she published the monograph *Ten Lectures on Wavelets,* which caught the attention of the American Mathematical Society. In 1994 it awarded her the Steele Prize for Mathematical Exposition, citing her groundbreaking work in wavelet theory:

> *Part of the accomplishment of Daubechies is finding those places where the concept arose and showing how all the approaches relate to one another. The use of wavelets as an analytical tool is like Fourier analysis—simple and yet very powerful. In fact, wavelets are an extension of Fourier analysis to the case of localization in both frequency and space. And like Fourier analysis, it has both a theoretical side and practical importance.*[5]

Many honors followed. In addition to earning the title of baroness (an award from King Albert II of Belgium), in 2000 Daubechies became the first woman recipient of the National Academy of Sciences Award in Mathematics. She's also an elected member of the American Academy of Arts and Sciences, the United States National Academy of Sciences, and the National Academy of Engineering.

In 1994, Daubechies became the first female full professor of mathematics at Princeton University, where she was active within the Program in Applied and Computational Mathematics until 2010. Since 2011 she has been sharing her expertise with

Below: Daubechies, the first woman to be elected president of the International Mathematical Union in 2011, holds the key to the union's new office in Berlin, Germany.

Above: A ca. 1903 photograph of Nassau Hall at Princeton University, where Daubechies became the first female full professor of mathematics.

students at Duke University, teaching courses in mathematics as well as electrical and computer engineering. During the Duke Summer Workshop in Mathematics—a program Daubechies founded with colleague Heekyoung Hahn—she shares her love of mathematics with intelligent and ambitious female high-school students.

Opposite: The door of the National Academy of Sciences building in Washington, D.C. shows an engraving of the ancient Greek mathematician Euclid. Daubechies made history in 2000 when she became the first woman to receive the organization's mathematics award.

Mathematical Crochet

DAINA TAIMINA

b. August 19, 1954

"Why should I trust something I cannot imagine?"[1]

—Daina Taimina, 2010

Above: Daina Taimina holding one of her spectacular crocheted hyperbolic plane models.

From a very young age, Latvian-born Daina Taimina excelled at mathematics, often understanding concepts faster than her peers and trying the patience of her instructors when she wanted to move on before her classmates were ready. In fact, in an interview with the *New York Times* Taimina reveals that a teacher was forced to send home a note to her parents after her rather vocal "I can't stand these idiots!"[2] in the second grade after her peers failed to understand a math lesson. Inspired by a high-school teacher who had a fun and engaging personality that kept more advanced students interested in the class, Taimina decided that this approach was how she would inspire her own students one day if she were given the chance.

In pursuit of that goal, she enrolled in the University of Latvia, studying mathematics and theoretical computer science. In 1977 she graduated summa cum laude with a master's degree in physics and mathematics, and then she taught college and high-school mathematics. In 1990 she obtained her PhD equivalent degree from the Institute of Mathematics, Academy of Sciences of Belorussia, after completing her dissertation, "Behavior of Different Types of Automata and Turing Machines on Infinite Words." When Latvia declared independence from the Soviet Union a year later, Taimina was able to get her formal PhD in mathematics from the University of Latvia, where she served as lecturer for more than two decades. Taimina also worked as an editor in a publishing house in Latvia before moving to the United States in the 1990s.

"I never knew that there is this 'girls and math' issue until I came to United States," says Taimina. "I graduated from the best school in Latvia and had wonderful teachers."

She believes that the idea of women not being good at math is a carefully formed construct. "The question 'women and math' was artificially created in the West not to undermine male supremacy and to keep women at home."[3]

After joining the mathematics department at Cornell University, where her husband, David Henderson, was a math professor, Taimina turned her crochet hobby into a groundbreaking idea. Instead of just imagining the spaces in non-Euclidian geometry, Taimina wanted to construct something tangible that represented the field's complex ideas. Hyperbolic planes had always been conceptual models, like the popular Poincaré disk model. While on a camping trip in 1997, however, she began crocheting a chain that she surmised might turn into a hyperbolic model that could maintain its original shape. With precise planning and tight stiches, she crocheted a surface to represent a hyperbolic plane. The result? Taimina successfully created the first physical model of a hyperbolic plane—one that you didn't have to imagine, but could actually see and touch.

Taimina can't keep track of how many hyperbolic crochet models she's created over the years for schools across the country, but she has inspired thousands more to create their own models. "They keep enjoying them even now after twenty years," she says. While Taimina created many of her crochet

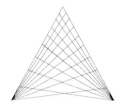

WHAT IS A HYPERBOLIC PLANE?

For centuries, mathematicians believed that Euclidean geometry (the study of plane and solid figures) was the only kind of geometry that existed. Euclid's five principles, or "axioms," were always accepted as truth and used to prove numerous mathematical theorems. It was the fifth axiom, known as the "parallel postulate," that mathematicians disproved, resulting in the notion of non-Euclidean geometry. Hyperbolic geometry is one branch that looks closely at the complexities of curved surfaces like the dramatic curves of a coral reef or a lettuce leaf (see more about hyberbolic geometry on page 182). As Cornell professor David Henderson explained in an interview with *Cabinet* magazine, one way to understand a hyperbolic plane "is that it's the geometric opposite of the sphere. On a sphere, the surface curves in on itself and is closed. A hyperbolic plane is a surface in which the space curves away from itself at every point."[4] Similarly, Daina Taimina explains the difference between elliptic and hyperbolic non-Euclidean geometries by asking us to imagine an orange and a piece of curvy lettuce. Flat 2D objects have "zero curvature," whereas the surface of an orange has "positive curvature" and a piece of curvy lettuce has "negative curvature."

But hyperbolic geometry isn't just an abstract concept; it has plenty of real-world applications today, especially in computer animation. There are also examples of hyperbolic

geometry in nature, including the frilly mantles of sea slugs and the irregular surfaces of cancer cells. And for many professionals—architects, surgeons, engineers, and more—knowledge of hyperbolic space is useful. For example, plastic surgeons must understand the way in which skin grows around wounds in order to keep scarring to a minimum. Having a physical model like Taimina's that students can hold and examine helps them better understand complicated, abstract mathematical theories and keeps them engaged, in the same way that her high-school teacher kept her engaged years earlier.

Above: Sea Slug

masterpieces to illustrate a classroom concept, they are also valued for their aesthetics. Her models, which have taken anywhere from ten hours to eight months to complete, have been highlighted in art and hobby magazines and have been displayed in exhibitions, art galleries, and museums, including the American History Museum of the Smithsonian Institution in Washington, D.C. In 2009, she wrote a book, *Crocheting Adventures with Hyperbolic Planes*, which was selected by the Mathematical Association of America to receive the 2012 Euler Prize. She just finished working on the second edition, which includes brand-new models. Her talks are just as popular among mathematicians as among artists, who appreciate the color and detail of her work. Combining her love of math education, knitting, crocheting, and art has allowed Taimina to reach outside the classroom and beyond imagination.

Above and Left: Samples of Daina Taimina's world-famous crochet models of hyperbolic planes.

TATIANA TORO

b. July 5, 1964

From Colombia to California, She Refused to Take No for an Answer

---

"Persevere no matter what. If you have a dream, let it guide you even if it looks impossible or crazy."

—Tatiana Toro

Born and raised in Colombia, Tatiana Toro realized she loved math as an elementary-school student when a counting lesson in the schoolyard grew into something more meaningful. "I was in first grade when we were learning to count, and we had these beautiful sets of blocks. Triangles, circles, squares . . . and they were all different colors. This crazy teacher would take us outside to draw these big sets of blocks on the ground with chalk. We were drawing structures like buildings, and it was really that we were learning set theory. My classmates weren't getting it, but I thought it was such fun."

WHAT IS SET THEORY?

Set theory is an essential component of all branches of mathematics. At its most basic, a set is a collection of things, and they are described within the set brackets like these: {x, y, z}. Each item within a set is called an element. To illustrate a set of states that begin with the letter "a," one could write:

{ALABAMA, ARKANSAS, ARIZONA, ALASKA}

The elements of the set are the states that begin with the letter a. Order does not matter within a set. If there is a set that does not have any elements within it, it is called a "null" set and is described with empty brackets like this:

$$\{\ \}$$

All mathematical objects can be combined into various sets, which is why sets are so helpful in all branches of mathematics, and especially in fields like probability and statistics.

When Toro was seventeen years old, she learned that Colombia was being invited to one of the most prestigious mathematical competitions in the world—the International Mathematical Olympiad—for the first time. Her school had been chosen to select candidates. Toro was in advanced math classes, and she knew she qualified. "I was a junior at the time, and I asked the teacher, 'Can I go and represent the school at the Olympiad?'" The teacher told her that four people had already been selected—all boys. She continued to press, and the teacher told her that she could go if she found space in the car, but she could not represent the school. She remained undaunted. "I found a place in the car and went with them to the Olympiad. Once I got there, I asked the officials, 'Can I please stay?' They let me stay, and I got a position on the team to represent Colombia and the school." Toro competed successfully in the 1981 International Mathematical Olympiad and went on to earn a bachelor's degree in mathematics from the National University of Colombia. Toro's advice to others is based on her experience. It is essential, she says, to "follow your dream no matter what . . . even if they say you can't."

THE INTERNATIONAL MATHEMATICAL OLYMPIAD: NO CALCULATORS ALLOWED

The prestigious International Mathematical Olympiad is an annual mathematical competition for high-school students all over the world. First held in Romania in 1959 with a mere seven participating countries, it has now expanded to more than one hundred countries representing five continents. Contestants are given six problems, which they need to solve over a two-day period. China has yielded the highest number of winning contestants with the Soviet Union following close behind. The US and Hungary are currently tied for third place. Women were once rare at the competition (some countries like Bulgaria have sent more than twenty women to the Olympiad from 1959 to 2008, while other countries like the US didn't send a single female until 1998), but the numbers have very slowly been increasing every year (see the graph below from the International Math Olympiad).

(A FEW) MORE GIRLS COMPETING AT IMO

Average number of girls on each team since 1979*

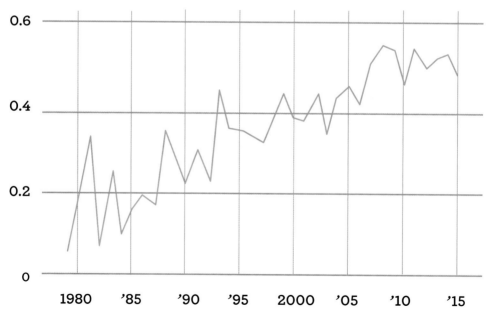

*The tournament has continously had at least 20 teams since 1979.

When Toro told friends and family in Colombia that she wanted to pursue a career in mathematics, her decision was met with skepticism. "Everyone told me it was crazy. I was not going to be able to make a living." But she applied to graduate schools in America, including the prestigious Stanford University, which has a notoriously selective program. She got in and in 1992 received her PhD from Stanford under the supervision of award-winning mathematician Leon Simon.

> *"Everyone told me it was crazy. I was not going to be able to make a living."*
>
> —Tatiana Toro

One of the main lessons she learned during her time at Stanford was the need to go back to the beginning: to return to the original hypothesis. She says, "You play a game with a set of rules. Let's say I play a game for 3 months. I build a strategy in that time . . . a winning strategy. I might want to show you that strategy, but I need one more thing to prove it to you. If I cannot do that, I go back to the beginning. Back to the original set of rules for the game. The solution you're looking for is usually at the beginning."

After accepting positions at the Institute for Advanced Study, University of California, Berkeley, and later at the University of Chicago, Toro joined the University of Washington faculty in 1996 to serve as the Craig McKibben and Sarah Merner Professor in Mathematics. From 2012 to 2016 she served as the Robert R. and Elaine F. Phelps Professor in Mathematics. Toro then headed back to UC Berkeley to become the Chancellor's Visiting Professor in Mathematics while participating at the Mathematical Sciences Research Institute (MSRI) in the Harmonic Analysis program. She served as a bridge between graduate students in the department and researchers visiting MSRI and taught a graduate course on recent developments lying at the interface of harmonic analysis, partial differential equations, and geometric measure theory.

Toro became a Guggenheim Fellow in 2015 and was elected as a member of the 2017 class of fellows of the American Mathematical Society "for contributions to geometric measure theory, potential theory, and free boundary theory." Toro's research explores the fact that some objects, viewed macroscopically, may appear very intricate, and when looked at microscopically, they may reveal hidden patterns or structures. It is these hidden patterns and structures that are used to provide detailed descriptions of the initial objects (see opposite). Toro has been recognized with a Simons Foundation Fellowship, an Alfred P. Sloan Research Fellowship, and a National Science Foundation Mathematical Sciences Postdoctoral Research Fellowship.

Toro is a firm believer in diversity in the mathematical sciences. She coorganized the first Latinos in the Mathematical Sciences (Lat@math) conference at UCLA in 2015. She is a leader in the movement to help Latino men and women pursue careers in these fields. The annual event promotes the advancement of Latina/o individuals currently in the discipline, showcases research being conducted by those at the forefront of their fields, and helps build a community around shared academic interests. As a minority in the field, Toro knows the importance of support and encouragement for success in math. She credits her mother and grandmother for her strength as she pursued her dreams. "I come from a country that is very male-oriented, so I know had it not been for my mother and grandmother believing in me, I wouldn't be where I am today."

KAREN E. SMITH

b. May 9, 1965

An International Expert in Commutative Algebra

"Instead of looking around your graduate program and worrying about how many students are 'better' than you, why not look around for someone you can help pull up?"[1]

—Karen E. Smith, 2017

As a sophomore at Princeton, Karen E. Smith remembers the thrill she felt when she read about the founder of modern algebra and saw the pronoun "she." That "she" was Emmy Noether (see page 36). Smith says everything changed after that moment.

"Probably until that point, I would have said that I never felt constrained by gender, but wow," she said in a 2017 interview. "In a way, that is hard to explain to those privileged enough to never, consciously or unconsciously, question whether or not they belong."[2] Smith hopes that young women today will feel that they do belong and are important.

Smith was born in Red Bank, New Jersey, near the shore. She enjoyed mathematics from a young age, and during middle school she explored modular arithmetic. She also read math books for recreation. One in particular—about Fibonacci numbers—caught her attention, and she read it multiple times. When Smith was a high-school senior, a calculus teacher offered an extra math class on number theory, basing the curriculum on a book by the mathematician and popular writer Underwood Dudley. Smith loved it.

During her first year at Princeton University, Smith's talent and enthusiasm for mathematics got the attention of former child prodigy and calculus professor Charles Fefferman. He suggested to Smith that she could make mathematics a career. Such an idea had never occurred

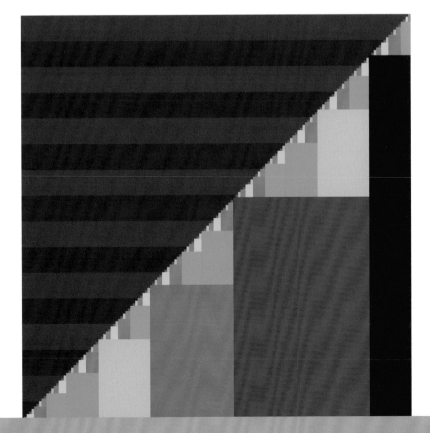

Left: Belgian mathematician Edouard Zeckendorf's theorem is illustrated by this colorful representation of the first 160 natural numbers as sums of Fibonacci numbers.

to her, but—to her parents' dismay—she promptly decided to switch her major from engineering to mathematics. Smith graduated from Princeton in 1987 as one of a small cohort of math majors, earned a teaching credential, and began teaching high school mathematics in New Jersey public schools. She became discouraged by the lack of support for young teachers, however, and decided to follow the graduate-school path after hearing about a friend's experience as a doctoral student.

In 1988, Smith became a graduate student instructor at the University of Michigan, where she met assistant professor Juha Heinonen, a Finnish mathematician. They married in 1991. Still, it wasn't until she reached her fourth year of graduate school that she began to seriously envision herself as a professional mathematician. "I had many jobs before mathematician, including lifeguard, hotel maid, deli meat slicer, computer parts recycling factory worker, pizza delivery person, SAT prep course instructor, and high school teacher," she says. "Of course, I had a few more typical jobs as well, such as babysitter, math tutor, and camp counselor. I went to a fancy college, but I have more modest working class roots. I know how to hustle to make a buck!"[3]

In 1993, Smith submitted her dissertation "Tight Closure of Parameter Ideals and F-Rationality" and earned her PhD in commutative algebra under the direction of Mel Hochster, an encouraging and supportive advisor. That same year, she accepted a National Science Foundation postdoctoral fellowship at Purdue University under the direction of Craig Huneke, who codeveloped the theory of tight closure with Hochster in the 1980s.

Smith's study of algebraic geometry flourished when she moved to Boston to accept a Moore Instructorship at Massachusetts Institute of Technology, where she was promoted to assistant professor. With colleagues and collaborators, she laid the foundations of a successful research program aimed at proving theorems about singularities[4] and vanishing of cohomology.[5]

In 1996, Smith received a National Science Foundation's prestigious CAREER Award for her proposal "Interactions of Commutative Algebras with Analysis, Geometry and Computer Science." In it, she stated her goal of establishing a computer-based graduate course in algebraic geometry at the University of Michigan, as well as creating a junior-level seminar focused on projective geometry for math and computer-science majors. After returning to Ann Arbor the following

year, she resumed her research in commutative algebra and algebraic geometry.[6] By 2001, Smith had established herself as a world leader in the study of tight closure, and, as a result, she received the 2001 Ruth Lyttle Satter Prize, awarded every two years to recognize an outstanding contribution to mathematics research by a woman.

In addition to her research, Smith has been an editor for several mathematics journals, including the *American Journal of Mathematics, Advances in Mathematics*, the *Journal of the American Mathematical Society*, and *Annales de Toulouse*. As Keeler Professor of Mathematics at the University of Michigan, she has trained numerous PhD students and postdoctoral scholars, many of whom are now outstanding researchers. Currently she is the faculty advisor for the local student chapter of the Association for Women in Mathematics. Smith mentors many of the university's African American math majors and has a special interest in advising those who transfer to the school from community colleges.

Smith is also a visiting professor at the University of Jyväskylä in Finland, where she has a special connection: it was the homeland of her husband, who passed away from kidney cancer in 2007. The couple's three children include a daughter, Sanelma, born in 1998, and boy-girl twins, Tapio and Helena, born in 2003.

Above: Inside the **Alchemist** *sculpture by Jaume Plensa, dedicated to the many researchers, scientists, and mathematicians who have studied at MIT, including Smith.*

Thirty years after being inspired by the knowledge that the founder of modern algebra was female, Smith was invited by the Association for Women in Mathematics and American Mathematical Society to give the Emmy Noether Lecture[7]—a nod to Smith's important contributions to the mathematical sciences and to her efforts to ensure that talented mathematicians are welcomed and supported.

WHAT IS TIGHT CLOSURE?

In order to understand the theory of tight closure, one first has to understand the branch of mathematics to which it belongs: commutative algebra. Basically, in contrast to noncommutative algebra (see page 190), the order of operations doesn't affect the result in commutative algebra (as is the case with simple multiplication and addition). This type of algebra examines commutative rings—algebraic structures involving a set (S) and the addition and multiplication binary operators. Ideals are special subsets (e.g., multiples of 3) of rings, and these ideals have "closure" if a member of the same subset is always produced by a given operation. For example, positive ideals are considered "closed" under the operation of addition, but not subtraction, because adding positive numbers always produces positive numbers, whereas subtracting positive numbers from one another does not.

Tight closure involves ideals of the positive prime ($p > 0$) characteristic in commutative Noetherian rings (see page 40). Smith defines tight closure as follows:

Let R be a Noetherian domain of prime characteristic p, and let I be an ideal with generators ($y_1, \ldots, y_r$). An element z is defined to be in the tight closure I^* if there exists a non-zero element c of R such that

$$(*) \quad cz^{p^e} \in (y_1^{p^e}, \ldots, y_r^{p^e}) \text{ for all } e \gg 0.[8])$$

The $\in$ symbol basically translates to "is an element of" and the $\gg$ symbol translates to "much larger than."

GIGLIOLA STAFFILANI

b. March 24, 1966

The Only Female Professor of Pure Mathematics
at Massachusetts Institute of Technology (MIT)

"I don't think it's productive to sit around and cry about
[the lack of women in the field]. That doesn't solve anything.
You prove people wrong by solving difficult theorems."[1]

—Gigliola Staffilani, 2008

Gigliola Staffilani credits good fortune for her success—"I have so many instances where I can zoom in and see the luck. My life should have gone any other way than the way it did."[2] However, she has truly persevered through hard work and determination.

Staffilani grew up in the rural town of Martinsicuro, located in the Abruzzo region of Italy, along the Adriatic Coast. She shared a "very crowded"[3] farmhouse with her parents, an older brother, uncle, aunt, two cousins, and two grandparents. She and a friend next door would spend countless afternoons exploring, helping with chores, making toys, and inventing new ways to entertain each other. "I think this was a great way to grow up,"[4] she says.

No one in Staffilani's family had gone to school past the fifth grade until her brother, ten years her senior, graduated from high school and went on to become a physician. For many years, there were no books in the house other than her and her brother's school books. When her father invested in a set of illustrated encyclopedias that had "a bit of everything,"[5] and her brother subscribed to the Italian version of *Scientific American* magazine, Staffilani spent many hours reading, which opened her mind. She says, "Of course I couldn't understand much, but it was in these journals that I learned about places such as MIT, which, for me at that time, was as far as Mars."[6]

In Italy in the 1970s, the school curriculum was rigorous. Staffilani recalls, "In every subject you were expected to do much more than the average student could accomplish, so we were challenged at every step."[7] She performed well in math and was well respected by teachers and peers. At the age of ten, when her father died of colon cancer, it was math that helped her deal with the tragic loss. She spent hours solving problems that were not assigned as homework, making her an even better and more confident math student. She became interested in science—a world without emotions: "I had too much of that in my family already,"[8] she says. When it was time to attend high school, she selected one with a reputation for being demanding, especially in math. There, Staffilani had a math teacher who encouraged her studies, challenged her with difficult math problems, and helped convince her mother to allow her to attend college. Due to the family's poor economic situation, Staffilani had been expected to become a hairstylist and marry one of her brother's physician friends. "My math teacher and my brother convinced my mother that a math degree would allow me to become math teacher in a high school, a very respectful job that can be easily combined with the care for my family,"[9] she says.

> *"Being smart was and still is not cool, both in the USA and Italy. I didn't care about that. I was never into girly things anyway."* [10]
>
> —Gigliola Staffilani

Above: A bird's-eye view of Bologna, Italy, where Staffilani attended college.

Staffilani received financial aid to attend the Università di Bologna, but it didn't cover living costs. She lived in the hallway of a palace that nuns ran, cooking on a camping stove. When she learned from an American doctoral student that she could become a professional mathematician, she applied for a PhD program in the United States and, thanks to a scholarship, left Italy immediately after college to continue her math studies at the University of Chicago. It was a challenging time for Staffilani because she didn't pass the TOEFL (Test of English Foreign Language) and was not going to be admitted to the school. After two weeks of hanging around the math department, broke and ready to return home to Italy, she was admitted, but her financial-aid check was delayed. Paul Sally, a well-known mathematician at the University of Chicago, gave her a check for $1,000, saying that she was a "true talent" who had "arrived at the right place all by herself." Exempted from the TOEFL requirement, Staffilani was determined to prove that she belonged in the program: "As long as it was formulas on a blackboard, I could survive."[11]

Staffilani earned her PhD in 1995 and went on to study difficult mathematics at the Institute for Advanced Study, which led to a Szegö Assistant Professorship at Stanford University. She was tenured at both Stanford University and Brown University (and taught at Princeton University as well) before accepting a position at MIT in

2002 as tenured associate professor. She became a full professor in 2006. Since 2007, she has been the Abby Rockefeller Mauzé Professor of Mathematics and served as associate department head of the department of mathematics from 2013 to 2015.

Staffilani is an analyst, with a concentration on dispersive nonlinear partial differential equations (PDEs). She is a pure mathematician working on certain partial differential equations that have been proposed by physicists as a model for a variety of wave phenomena, such as the behaviors of a diluted gas at extremely low temperatures, the behavior of surface waves in an ocean or in a shallow channel, and the interaction of galaxies. She uses "very sophisticated mathematical tools to learn about their properties and how they interact with each other in a very complicated way" and describes her work as follows:

The equations are nonlinear, which means that their solutions cannot be summed in order to get another solution. Quite the opposite the nonlinearity is such that understanding what happens when one interacts multiple waves solutions is an extremely challenging problem. It is in fact to study this interaction that mathematicians such as myself are called into the game. The tools that I use in my research are harmonic and nonlinear Fourier analysis, number theory, dynamical systems, differential geometry, and probability.

I love the intellectual challenges of the problems I work on because to solve them I need to keep a very open mind and believe that at the right moment the perfect tools may come from an inspected area of mathematics. I never get bored! [12]

At MIT, Staffilani has taught differential equations and multivariable calculus and has had the opportunity to create her own classes. In addition to earning teaching awards and prominent fellowships, Staffilani is an American Mathematical Society fellow and a member of the American Academy of Arts and Sciences. In 2017, she was awarded a prestigious Guggenheim Fellowship. She also organizes Women in Math conferences and, whenever possible, extolls the virtues of being a mathematician. "When I meet a student who loves math, I always tell him or her that I have a wonderful profession," she said. "Being able to direct your own thinking, to explore what you want to explore and to explain what you achieved to other people who are interested is quite a dream job."[13]

As focused as she is as a pure mathematician, Staffilani values her roles outside academia as well. She is married to fellow MIT mathematics professor Tomasz Mrowka and the couple has two children: Mario and Sofia. "Having a life outside of mathematics—I am also a mother and wife—helps too. It balances other forces that constantly act on me: being a teacher, researcher, mentor, administrator, and so on."[14]

Like the dispersive equations she studies, Staffilani's life has taken an unpredictable path. With her love of difficult problems and ability to face down any obstacle that comes her way, she is well positioned to keep breaking boundaries in math and in life.

Below: Staffilani in her office at MIT.

INTERCONNECTIONS

"Tatiana Toro [page 138] is a friend of mine from the time I was a graduate student at the University of Chicago, and she was an assistant professor there too. I served with Maryam Mirzakhani [page 178] on the Scientific Advisory Committee at MSRI for three years. She was a wonderful mathematician, of course, but she was also a wonderful person, very thoughtful and very calm. I have interacted a few times with Ingrid Daubechies [page 126] at conferences or other events. I admire her greatly, and when I was at the beginning of my career, she was definitely my hero. Her work on wavelets was very influential for me. I met Karen Smith [page 144] for the first time during the summer mentoring program at Princeton in 1995. She was already a rising star in her field, but that was not what stayed with me. She is an incredibly funny and deep person, all at the same time. I interacted with Chelsea Walton [page 186] while she was a Moore Instructor at MIT few years ago. She is a remarkable person, capable of juggling the intense demands of a career as a researcher and her outreach frequent engagements."[15]

ERICA N. WALKER

b. November 19, 1971

An Advocate for Advanced
Mathematics Education for the Underserved

"When I walk through city neighborhoods or through the halls of urban schools I do not see them as empty of promise . . . I look for the talent that I know is there."[1]

—Erica N. Walker

E rica N. Walker is a mathematics educator who is "interested in understanding how people learn and do mathematics throughout their lives, and exploring how schools and communities can facilitate better mathematics learning and engagement for everyone, from little kids to adults."[2]

For Walker, her overall love of learning, along with her math education, was naturally and enthusiastically encouraged by the neighborhood she lived in, where education was endorsed and supported: "So I always thought that I could be or do anything I wanted," says Walker. "That kind of intellectual confidence is priceless."[3] From a young age, learning was a pleasure for Walker. Throughout her childhood, she was encouraged by her parents, teachers, neighbors, and peers. About these early cheerleaders, she writes, "All of these people instilled in me a love of learning and underscored daily that learning is something to be enjoyed, and it is something you never stop doing."[4] Because of Walker's upbringing, she has dedicated her career to advocating for better math education at the high-school level, especially for underserved populations.

Walker majored in mathematics at Birmingham-Southern College (graduating cum laude) and then attended Wake Forest University to obtain a master's degree in mathematics education, becoming certified to teach high-school mathematics, which she did in DeKalb County, Georgia. During this time, she developed the goal of encouraging more African American students to take advanced math classes to help prepare them for upper-level college math courses. Her advice to students is to pursue advanced classes. "Advanced math is critical," she says, "because it increases your chances of going to college. There are more career options open to you."[5]

Above: Walker is inspired by bridges, especially the Brooklyn Bridge here, as feats of creativity, imagination, mathematics, and engineering.

"As a mathematics educator, I'm interested in understanding how people learn and do mathematics throughout their lives, and exploring how schools and communities can facilitate better mathematics learning and engagement for everyone, from little kids to adults."[6]

—Erica N. Walker

The impact of advanced math coursework in high school became the focus of the dissertation Walker wrote for the PhD she received in 2001 from the Harvard Graduate School of Education. "On Time and Off Track: Advanced Mathematics Course-Taking Among High School Students" examined students' experiences taking math classes and analyzed outcomes, including who continued on and who stopped. Walker became a postdoctoral fellow at Teachers College, Columbia University, and joined the faculty in 2002.

Walker is now a full professor of mathematics education at Teachers College, Columbia University, and describes her research as follows:

> *In my work I draw on history and contemporary contexts to explore how people learn and are socialized to do mathematics in various "communities" (peer, neighborhood, school, family and home). I also explore how teachers can capitalize on students' strengths (rather than focus on students' "deficits") to improve students' mathematics outcomes (performance, participation, and persistence).[7]*

Walker's work has been published in various peer-reviewed journals, such as the *American Education Research Journal*, the *Journal for Urban Mathematics Education,* and the *Urban Review*.

WHAT ERICA N. WALKER'S RESEARCH HAS ACCOMPLISHED

"My work has sought to help move the field of mathematics education forward. So much research previously focused on documenting Black and Latino/a students' underperformance. By focusing on the experiences—in and out of school—of successful students, I uncovered powerful academic communities and rich spaces for learning and doing mathematics in meaningful ways that could be leveraged for struggling students as well."[8]

She is also the author of two books: *Building Mathematics Learning Communities: Improving Outcomes in Urban High Schools* (Teachers College Press, 2012) and *Beyond Banneker: Black Mathematicians and the Paths to Excellence* (SUNY Press, 2014). Her later book includes interviews with thirty mathematicians, including Dr. Evelyn Granville, the second African American woman to obtain her PhD in mathematics from an American university. "I think about many of ['the mathematicians'] stories [in *Beyond Banneker*] and how their experiences can help us better understand how to inspire young people's mathematics interest, both in and out of school."[9]

In addition to her academic work, Walker consults with nonprofit organizations, schools, and teachers. In a 2016 interview in *Scientific American*, Walker responded with this to a question about supporting diversity in the field:

The broader mathematics community should be committed to ensuring that there is equal access to mathematics in our schools. For example, at the secondary level, I am really concerned about the limited availability of advanced math courses in certain schools. I think mathematicians and mathematics educators alike should really be vocal about this inequity, and active in doing something about it.[10]

TRACHETTE JACKSON

b. July 24, 1972

Curing Cancer with Mathematics

"Some people will look right through you and some will deliberately turn a deaf ear to you. Do not be discouraged— be visible, be present, be heard!"[1]

—Trachette Jackson, 2017

When battling a complex disease like cancer, a strong defense is imperative. Trachette Jackson, professor of mathematics at the University of Michigan, has joined the fight, wielding mathematical oncology as her weapon of choice. It's an exciting time to be in the field, she says.

"Mathematical and computational modeling approaches have been applied to every aspect of tumor growth from mutation acquisition and tumorigenesis to metastasis and treatment response," says Jackson. "My research focuses on developing mathematical approaches that are able to address critical questions associated with vascular tumor progression and targeted therapeutics."[2]

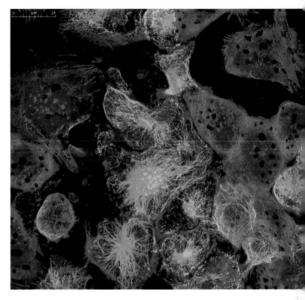

Below: Microscopic tumor cells labeled with fluorescent molecules.

Understanding how cancer develops is key to figuring out effective ways to treat it. A substantial amount of current research into improving cancer prognosis focuses on finding ways to selectively target pathways involved in the growth of tumors. Chemotherapy targeted at shrinking tumors and preventing metastasis is a standard and moderately effective treatment, but its potential to damage healthy tissue, suppress the immune system, and inflict untold suffering on the patient means that it comes at a high cost. For this reason, Jackson and many other researchers around the globe are searching for connections between cancer proliferation and other changes in the body, including changes to the vascular system. As Jackson explained in a 2013 interview with *Scientific American*,

In the last few years, how tumors initiate blood vessel formation has become a big topic. We are looking at the mechanistic aspects of blood vessel formation in response to tumors. We're asking questions about how the biomechanics and biochemistry connect in order to give this strange conglomeration of vessels . . . If you could target something like blood vessel formation, it would hopefully have fewer side effects [than chemotherapy].[3]

Born in Monroe, Louisiana, Jackson moved every two years to a new state or country, following her father's assignments in the US Air Force. When she was twelve years old, the family settled in Mesa, Arizona, where Jackson attended a large public high school beginning in 1987. She spent the summer after her junior year at the

SO, HOW DO YOU
FIGHT CANCER WITH MATHEMATICS?

Jackson's mathematical skills come into play in this research because of the suite of mathematical models she and her collaborators developed that evaluate the biochemistry of tumor growth, vascular composition, and therapeutic outcomes. "We use these models to investigate the integrated effects of various molecular players on the bidirectional communication (i.e. crosstalk) between endothelial cells and tumor cells that contribute to and enhance key aspects of tumorigenesis," says Jackson.[4]

By learning more about these molecular mechanisms and basic pathways in the origination and development of cancers, doctors can develop cell-specific approaches to cancer therapy that operate at the molecular level. And here again, math comes to the rescue, as Jackson and her colleagues develop numerical models comparing the effects of different anticancer agents on intracellular molecules, including Bcl-2 proteins, over a period of days.

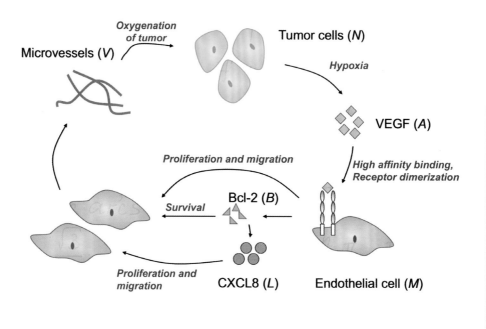

nearby Arizona State University's Math-Science Honors Program (MSHP), which brought together Phoenix-area minority students to learn about mathematics in an intensive, hands-on, and communal environment. It was here that her talent in calculus caught the attention of mathematics professor Joaquin Bustoz, Jr. In a recent interview, Jackson reflected on the experience which left a lasting impression on her:

> **The [MSHP]** *program taught me about commitment, self-discipline, and perseverance. It taught me to go confidently in the direction of my dreams. It helped me become familiar with the "hidden curriculum" that those of us who aren't "to the manor born" academically would have no way of knowing about.*[5]

Jackson graduated among the top twenty students of her approximately eight-hundred-student high-school class and went on to undergraduate studies at Arizona State University, originally intending to study engineering before Dr. Bustoz steered her toward a career mathematics. She switched from studying pure mathematics to mathematical biology after she attended a talk by her future

PhD advisor, Dr. James D. Murray, on the mathematics of "how the leopards got their spots."[6] In 1994, she received a Bachelor of Science degree in mathematics, and in 1996 she got her Master's degree at the University of Washington. Two years later she completed her PhD thesis, "Mathematical Models in Two-Step Cancer Chemotherapy," and earned her PhD in applied mathematics.

After graduate school, Jackson took postdoctoral positions at the Institute for Mathematics and Its Applications at the University of Minnesota and at Duke University before joining the faculty of the University of Michigan in 2000. Over the years, she has reviewed articles for the *Journal of Mathematical Biology* and the National Academy of Sciences and has published numerous papers on the subject of mathematical oncology. In 2003, she became only the second African American woman to ever receive the prestigious Alfred P. Sloan Research Award in Mathematics, and in 2005 the James S. McDonnell Foundation granted her the generous 21st Century Scientist Award for her research into mathematical models of the formation of new blood vessels associated with tumor growth. In 2006, she became coprincipal investigator of the University of Michigan's SUBMERGE (Supplying Undergraduate Biology and Mathematics Education Research Group Experiences) program, and in 2008 she became a full professor at the University of Michigan and served as the senior editor for the academic journal *Cancer Research*.

Currently Jackson codirects the Mathematics Biology Research Group (MBRG), an organization she cofounded that presents lectures, workshops, and other activities for graduate students and postdoctoral researchers in mathematical biology. She continues to receive honors and international attention for her mathematical contributions to the epic global battle against cancer and for serving as a role model and mentor for bright and ambitious young men and women of color. Looking back on her trials and accomplishments, Jackson has this advice for her younger self:

> *Don't be invisible! You are African American, you are a Woman, and you are an Applied Mathematician—being all three of these things can be powerful if you have the courage to fully embrace the essence of each and live your truth.*[7]

Above: In 1998, Jackson earned her PhD in Applied Mathematics at the University of Washington in Seattle.

CARLA COTWRIGHT-WILLIAMS

b. November 6, 1973

A Mathematician Using Her Skills
to Make a Difference in the World

"As long as you're alive, you can always have a new start.
I'm not really different from anyone else except for my
willingness to keep trying."[1]

—Carla Cotwright-Williams

South-Central Los Angeles has long been associated with crime and violence, and Carla Cotwright-Williams can attest to the neighborhood's checkered past. After all, she grew up there. "There were burglaries in the neighborhood. Gang violence was typical. My dad was a police officer, so he knew the spots to avoid," she said. "Our house was broken into several times, and they even stole our dog from our front lawn."[2]

The environment could have easily held her back, but her parents wouldn't let that happen. "My mother and father always made sure to emphasize the positive. Those things happened around me, but they didn't define me."[3]

By the time she reached middle school, her family had moved out of the true inner city and into a different neighborhood in LA, with better schools and more opportunity. "I tested into the accelerated classes at Westchester High School. We were already writing essays in ninth grade for college prep while other kids were just learning to write letters."[4]

Cotwright-Williams credits her maternal and paternal grandparents for putting such a strong emphasis on education, and says, "I knew that education was the ticket to my success."[5] Her decision to focus on a career in STEM was solidified in high school. During her sophomore and junior years, she participated in a summer enrichment program for underrepresented students hosted by the Minority Engineering Program. She attended courses in math and engineering at UCLA, learned from graduate students, and took field trips to NASA and Edwards Air Force Base. She went on to attend college at California State University, Long Beach, but encountered difficulties. "I was kicked out of college because of my grades," she says. "I had to work extremely hard to overcome the deficit in my GPA to regain my status as a regularly matriculating student."[6] She asked for help, made the necessary adjustments, and persevered to graduate with a BS in mathematics. She furthered her education and and received master of science degrees in mathematics from both Southern University and A&M College at Baton Rouge (2002) and the University of Mississippi (2004).

Cotwright-Williams started off as an engineering major in college and recalls the first time she met a black female PhD mathematician in graduate school, Dr. Stella Ashford: "She encouraged me to pursue a degree in pure mathematics," Cotwright-Williams says. "I never saw myself as a mathematician because of the many struggles I experienced as an undergrad."[7] Under Ashford's guidance, Cotwright-Williams completed a thesis on a branch of combinatorics called matroid theory and changed her focus from

> *"STEM fields are challenging for everyone. Don't compare yourself. I found that comparison can be your biggest enemy. I think comparing yourself to others leads to 'imposter syndrome' [see page 116] . . . Often, I've found our imaginations about the ease of success of others are false. So rather than focusing on others, focus on achieving your personal bests."*
>
> –Carla Cotwright-Williams

Above: Carla Cotwright–Williams, and her husband, fellow mathematician Bryan Williams.

math education to pure math. She received her PhD from the University of Mississippi in 2006. During this time, she met her future husband, Bryan Williams, also a mathematician. Though they work in the same field, Cotwright-Williams is careful to avoid working together in the same place. "I am my own person when it comes to my work,"[8] she said.

Cotwright-Williams held teaching positions at Wake Forest University and Hampton University before yearning for "something other than my tenure-track position."[9] She decided to look for other opportunities in the math world, and that is when she discovered there were math careers within the US government. But how do you make a career change of that magnitude? Instead of taking classes, Cotwright-Williams decided to join in faculty research opportunities with the federal government. "If I found a free education enrichment program, I signed up. Every time. I am convinced of the power of education."[10] She worked with NASA, studying the relationship between random graphs and Bayesian networks to improve methods for determining and maintaining systems health in autonomous avionics. With the US Navy, she worked with a team to examine statistical measures of uncertainty to create techniques for use with data-integrity problems. These research experiences became Cotwright-Williams's "bridge out of academia."[11]

Right: Being selected for the AMS Congressional Fellowship in 2012 brought Cotwright–Williams to Washington D.C., and it was a transformative experience for her. "Working on Capitol Hill was one of the greatest opportunities of my life! It was fast moving and you had be a quick learner."

She took graduate courses in public policy, participated in the EDGE Program (Enhancing Diversity in Graduate Education, see page 105) to develop a network of mentors and peers, and applied for the AMS Congressional Fellowship. She was not selected for the fellowship the first time she applied, so she waged a personal campaign to prepare herself before applying for it a second time. In 2012, she was awarded the fellowship, which allowed her to work on Capitol Hill as a science and technology fellow, bringing a new perspective to the halls of Congress. In 2013, she had the opportunity to be on-site with her committee's investigative team to speak with the first responders of the Boston Marathon bombing.

Her unique journey has allowed her to build a personal and professional network that includes the mathematics community and the broader scientific community, as well as the nonscientific, general public. She says, "Over my life, people have helped me. I figure I could never repay the help I received. I find it only responsible to help others in the same way."[12]

Below: First responders at the Boston Marathon Bombing in 2013.

In 2015, Cotwright-Williams became a Hardy-Apfel IT Fellow at the US Social Security Administration, which pays out benefits to over sixty million people, including retirees, children, widows, and widowers. Her work centers on using information (data) and math tools to gain insight into interesting aspects of government functions to help improve the services provided to US citizens. This work is just

WHAT IS A DATA SCIENTIST?

A phrase coined in 2008 by D.J. Patil and Jeff Hammerbacher, "data scientists" essentially work with large quantities of data or information from various places, link the information together, and make it available for analysis and interpretation. Big data companies, such as Facebook and LinkedIn, constantly collect data, and without data scientists, it would be difficult to interpret that information: Why are people clicking through one ad at a faster rate than others? Where are users finding your website? Why are some people browsing longer than others? So, if you're comfortable working with data and numbers, and have a knack for analysis, then the growing field of data science might be a career path to consider.

one of many ways the government's functions are improved through the use of analytics. "From national security to energy, data analytics can help drive informed decision-making,"[13] Cotwright-Williams says. She begins her work as a data scientist with the Department of Defense in 2018.

Cotwright-Williams's nontraditional mathematics career was influenced and supported by fellow mathematicians, family members, teachers, counselors, and others, she acknowledges. "They encouraged me to pursue my passion and explore careers where I might use both my analytic skills and my soft skills," she says. "It hasn't been the easiest thing. I have had to navigate uncharted waters and remarket myself for this new world.[14]

It's a challenging, complex world that requires her to take on many roles, including one that involves dismantling stereotypes. "Being the only mathematician in a room can provide a great deal of amusement when no one suspects you—I don't appear to be a 'typical-looking' mathematician," she says. "Expanding the perceptions of mathematicians is important to increasing our numbers and diversity."[15]

Networking helps facilitate this, Cotwright-Williams believes. "Just do it. Force yourself to do it. Take business cards and pass them out. It's a nice way to introduce yourself. I always have business cards in my pocket, whether I'm at a math conference or at church." She is a member of the BIG (Business, Industry, Government, and Academia) Math Network, which gathers like-minded individuals interested in careers in these fields. She also is a member of the public service organization Delta Sigma Theta Sorority, Inc., in which she introduces STEM-related programs to African American middle-school girls, some of them certain to become future mathematicians.

"It's easy to stick to the people you know, but when I'm out and about, I will introduce myself to a stranger in the room. You gotta just do it." [16]

—Carla Cotwright-Williams

Opposite: Carla Cotwright-Williams, data scientist for the U.S. Department of Defense.

EUGENIA CHENG

b. 1976

Ridding the World of Math Phobia

"For me, the most beautiful part [of mathematics] is the boundary between what we understand logically and what we don't. The more we understand, the more of that boundary we have, because the surface of the sphere grows."[1]

—Eugenia Cheng, 2017

The law of diminishing returns is a concept in economics that states that if one factor of production is increased while other factors are held constant, the output per unit of the variable factor will eventually diminish. Eugenia Cheng was taught this lesson, along with similar mathematical ideas, as a young child by her mother. She and her sister were supported and nurtured by both parents, who provided a strong foundation for Cheng's confidence, achievement, and desire to spread her knowledge in the most accessible—and edible—way possible.

Cheng was born and raised in England, where she attended all-girls schools until college. Girls played as they wished in school and specialized in academic subjects of the greatest interest to them. Many, including Cheng and her sister, chose math and science. Most of her teachers and other authority figures—including the prime minister and queen—were female. These factors in combination gave her the impression that she could do anything. Outside of her schooling, Cheng read voraciously, practiced piano, and ate everything in sight. For her undergraduate and postgraduate degrees (both in mathematics) she attended Gonville and Caius College of the University of Cambridge. Martin Hyland, a professor of mathematical logic, was her PhD advisor.

For her main research interest, Cheng chose higher-dimensional category theory, which she describes as the mathematics of mathematics. "Mathematics is about deeply understanding how the world works, and category theory is about deeply understanding how mathematics works," she says. "The more deeply you understand something, the better you can work with it."[2] True to her desire to make math understandable to a wider audience, Cheng wrote about category theory using analogies from one of her great interests: food. Each chapter of her 2015 book *How to Bake Pi* begins with a dessert

Opposite, bottom: Cheng's "Bach pie" combines banana, chocolate, and "mathematical knots" made of glazed pastry.

Left: Cheng attended Gonville and Caius College in Cambridge, England, where in 2002 she earned her PhD in pure mathematics.

recipe, which Cheng uses to explain concepts in mathematics, such as topology, groups and sets, and the process of proving theorems using axioms. The idea for the book came during a class she was teaching. One day a student asked her to explain a concept using Oreo cookies, so she did. "It was this thing called conjugation," she said in a 2017 interview with the *Guardian*, "where you multiply A by B and A inverse—you sandwich B between two As, one of which is the other way around. The cookie demonstrated that perfectly, because you have the cream filling between two cookies, but one of them is the other way around from the other."[3]

Cheng has taught at universities in Nice, France, and Sheffield, England, and she is currently honorary fellow of pure mathematics at the University of Sheffield. At the School of the Art Institute of Chicago, she teaches mathematics to arts students. Throughout her teaching career, she has used anecdotes so that students can relate mathematical topics to their lives. As she describes it:

I want to show everyone that math is fun and creative, that it is not just about getting the right answer, that it is about exploration and investigation, that it is not about memorizing formulae, that it is not just about numbers and equations, and that times tables are really not very important.[4]

Left: Cheng eating one of her "perfect" mince pies in front of several formulas that reveal how cooks can mathematcally determine the perfect balance between pastry and mince, November 2016.

Left: On November 14, 2015, Cheng joined Stephen Colbert on the Late Show for a cooking duel.

How to Bake π became a popular book, leading to many talks and appearances. On *The Late Show with Stephen Colbert* in 2015, Cheng and the host had a duel with rolling pins. Her talk "Mathematics and Lego: the untold story" and papers "On the Perfect Quantity of Cream for a Scone" and "On the Perfect Size for a Pizza" also present mathematical topics in a light-hearted manner. Cheng's latest book, *Beyond Infinity: An Expedition into the Outer Limits of Mathematics*, explores some of the field's most mind-boggling concepts.

Cheng, a professional pianist, says she uses her music to balance out her mathematical thinking. Music expresses and explores emotions, while mathematics is the language of logic. Her specialty is lieder—German poems set to classical music—which she has performed with singers and chamber musicians, including Ami Radunskaya (see page 120), at many recitals and events. Cheng was awarded the Sheila Mossman Memorial Award from the Associated Board of the Royal Schools of Music and was the first recipient of the Brighton and Hove Arts Council Award for the Musician of the Year. In 2013 she founded the nonprofit Liederstube in Chicago, where people present and enjoy classical music in an intimate, informal setting.

Next page: Cheng teaches about the connection between mathematics and music.

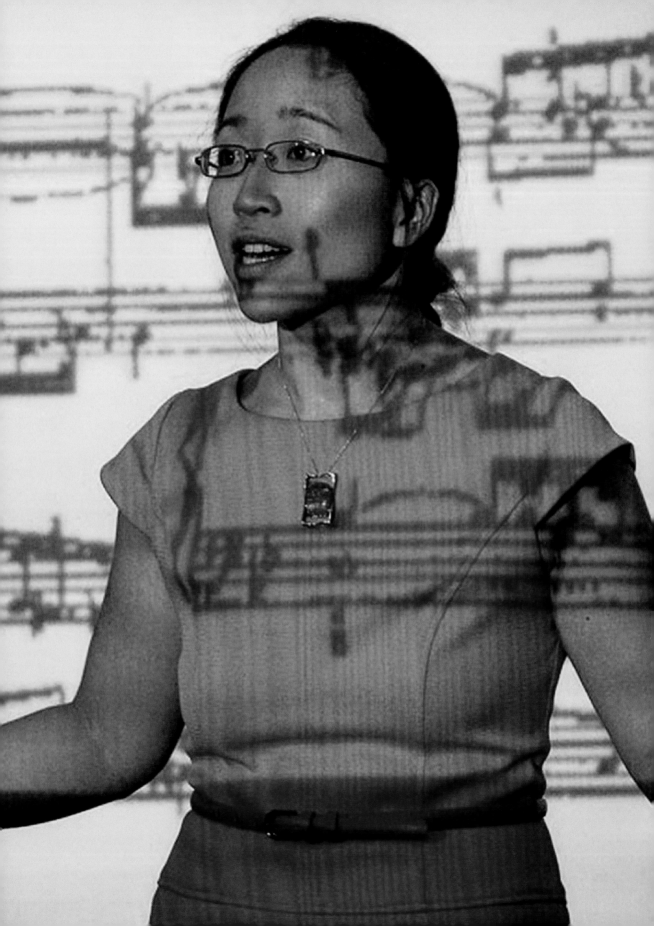

AN INFINITY
OF INFINITIES

One of the topics Cheng discusses in her new book is the concept of multiple infinities, which was initially put forth by the seventeenth-century German mathematician Georg Cantor. In addition to his invention of set theory, Cantor distinguished between several types of numbers, including cardinal numbers (i.e., natural numbers used for counting) and ordinal numbers (i.e., natural numbers used for ordering). His most controversial hypothesis introduced the idea of transfinite numbers, which refer to all numbers greater than finite numbers that aren't necessarily absolutely infinite. Cantor's counterintuitive hypothesis shook the mathematics community at the time, but Cheng found a useful metaphor with which to explain the mystery of infinity. In her words:

There is the thing about some infinities being bigger than others, but one of my favorite things is that one plus infinity is different from infinity plus one. It is like that Shakespeare thing of forever and a day—that forever and a day is longer than forever.[5]

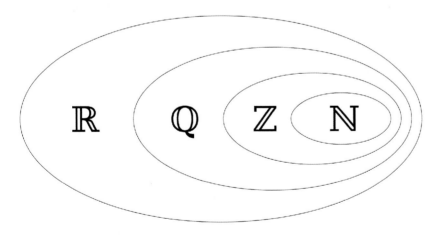

This diagram represents number system hierarchy, with real numbers (R) containing the ratonal numbers (Q), which contain the integers (Z), which contain natural numbers (N).

MARYAM MIRZAKHANI

May 3, 1977 – July 14, 2017

Only Female Recipient of the Fields Medal

"I like crossing the imaginary boundaries people set up between different fields—it's very refreshing."[1]

—Maryam Mirzakhani, 2014

Maryam Mirzakhani once told a reporter, "You have to spend some energy and effort to see the beauty of math."[2] And this she did, with the passion and fearless ambition of one whose brilliant light was extinguished all too soon.

Left: Mirzakhani received math's highest honor in 2014.

Born in Tehran, Iran, Maryam grew up during the Iran-Iraq war and its aftermath. She had ambitions of becoming a writer at first; she was a voracious reader and loved television biographies of famous women. In middle school she was a poor math student, at least according to one teacher. For a while she was discouraged from pursuing math, but her brother's strong enthusiasm for science was infectious:

Above: The Azadi Tower in Tehran, which commemorates the 2,500-year anniversary of the Persian Empire, was a frequent site of protests during the 1979 Iranian Revolution.

> *My parents were always very supportive and encouraging. It was important for them that we have meaningful and satisfying professions, but they didn't care as much about success and achievement. In many ways, it was a great environment for me, though these were hard times during the Iran-Iraq war. My older brother was the person who got me interested in science in general. He used to tell me what he learned in school. My first memory of mathematics is probably the time that he told me about the problem of adding numbers from 1 to 100. I think he had read in a popular science journal how Gauss solved this problem. The solution was quite fascinating for me.*[3]

Mirzakhani also credited her friendship with classmate Roya Beheshti with stoking her intellectual fire:

It is invaluable to have a friend who shares your interests, and helps you stay motivated. Our school was close to a street full of bookstores in Tehran. I remember how walking along this crowded street, and going to the bookstores, was so exciting for us. We couldn't skim through the books like people usually do here in a bookstore, so we would end up buying a lot of random books.[4]

Above: A view of Sharif University of Technology against the backdrop of the Alborz mountain range.

In the same 2008 interview, Mirzakhani recalled, "I never thought I would pursue mathematics until my last year in high school." At her all-girls school in Tehran, she asked her principal if she and Beheshti could take special mathematical problem-solving classes in order to compete for Iran's International Mathematical Olympiad Team. They were the first women in the competition's history to make the cut, and Mirzakhani gained international recognition in 1994 when she was awarded the competition's gold medal in Hong Kong. (Beheshti got the silver.) A year later Mirzakhani earned a perfect score and another gold medal in Toronto, Canada. That same year she went on to study mathematics at Iran's leading school for physical science, Sharif University. In 2008 she recalled the atmosphere of excitement and camaraderie at the school: "I met many inspiring mathematicians and friends at Sharif University. The more I spent time on mathematics, the more excited I became."[5]

Mirzakhani published several papers as an undergraduate and continued to compete in mathematical contests. A bus crash in 1998, however, almost extinguished her potential. In February of that year, she and other elite mathematicians were returning from a competition in the western city of Ahwaz when their bus skidded out of control and crashed into a ravine. Seven award-winning mathematicians and two drivers were killed, but Mirzakhani was one of lucky ones who survived. She later left Iran to study and work abroad.

After graduating with her bachelor of science degree in 1999, Mirzakhani was accepted to Harvard University, where Curtis McMullen, the 1998 Fields Medal winner, became her advisor. McMullen recalled Mirzakhani's "daring imagination."

She would formulate in her mind an imaginary picture of what must be going on, then come to my office and describe it. At the end, she would turn to me and say, "Is it right?" I was always very flattered that she thought I would know.[6]

A determined, relentless, and creative researcher, Mirzakhani quizzed her professors in English and made notes in her native Farsi. To solve problems, she made drawings accompanied by mathematical formulas on large sheets of white paper spread out on the floor—work her young daughter described as "painting." Mirzakhani likened the process to "being lost in a jungle": "[You try] to use all the knowledge that you can gather to come up with some new tricks, and, with some luck, you might find a way out."[7]

In addition to hyperbolic geometry, Mirzakhani's areas of expertise included moduli spaces, Teichmüller theory, ergodic theory, and symplectic geometry. Her work was highly theoretical in nature, but it informs quantum field theory and had secondary applications to engineering and material science. Within mathematics, her research had applications for the study of prime numbers and for cryptography.

In 2004 Mirzakhani was awarded her doctorate at the completion of her dissertation, "Simple Geodesics on Hyperbolic Surfaces and Volume of the Moduli Space of Curves." Considered a masterpiece, the work solved two longstanding problems about the volume of moduli spaces and how their measurement could be applied to string theory.[8] Either solution would have been newsworthy on its own, according to University of Chicago mathematician Benson Farb, but Mirzakhani connected the two into a thesis that yielded papers in each of the top three mathematics journals. It also earned her the Leonard M. and Eleanor B. Blumenthal Award for the Advancement of Research in Pure Mathematics.

WHAT IS
HYPERBOLIC GEOMETRY?

Mirzakhani was captivated by the geometric and dynamic complexities of curved surfaces,—hyperbolic surfaces—that is, surfaces that have a constant negative curvature, in contrast to standard Euclidean surfaces with no curvature and elliptic surfaces with a constant positive curvature.

Elliptic and hyperbolic geometries are considered "non-Euclidean," in that they do not conform to Euclid's parallel postulate, which states in essence that two lines are said to be parallel if they never intersect. To illustrate, in Euclidean geometry, two lines drawn perpendicular to another line at any point will remain at a steady distance from one another. However, in elliptic geometry these two lines will curve toward one another, and in hyperbolic geometry these two lines will curve away from one another.

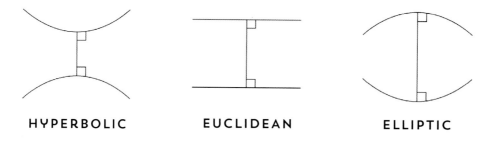

HYPERBOLIC   **EUCLIDEAN**   **ELLIPTIC**

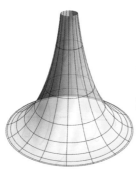

A classic example of hyperbolic geometry is the pseudosphere. Developed by the Italian mathematician Eugenio Beltrami in 1868, this surface has a constant negative curvature of $-1/R^2$, where R is the radius. (Compare this to a sphere with radius R and a constant positive curvature of $1/R^2$.)

Also in 2004 Mirzakhani accepted a Clay Research Fellowship and was appointed an assistant professor of mathematics at Princeton University. The fellowship allowed her to think about difficult problems, travel, and discuss ideas with other mathematicians, which suited her work style. "I am a slow thinker, and have to spend a lot of time before I can clean up my ideas and make progress," she said in a 2008 interview.[9]

In 2005 Mirzakhani married Jan Vondrák, a Czech theoretical computer scientist and applied mathematician at Princeton, and three years later the couple moved to Palo Alto, California, where Mirzakhani assumed a post as professor of mathematics at Stanford University. In 2011, their daughter, Anahita, was born.

Five years earlier, Mirzakhani had begun a collaboration with Alex Eskin at the University of Chicago to answer a mathematical challenge that physicists had struggled with for a century: the trajectory of a billiard ball around a polygonal table. Their two-hundred page paper, published in 2013, was hailed as "the beginning of a new era"[10] in mathematics. According to Stanford mathematician Alex Wright, "It's as if we were trying to log a redwood forest with a hatchet before, but now they've invented a chain saw."[11]

A year later Mirzakhani was awarded the Fields Medal for "her outstanding contributions to the dynamics and geometry of Riemann surfaces and their moduli spaces."[12] She was the first and remains the only female winner of the prize. At the time of the award, University of Wisconsin professor Jordan Ellenberg explained her research to a popular audience:

> [Her] work expertly blends dynamics with geometry. Among other things, she studies billiards. But now, in a move very characteristic of modern mathematics, it gets kind of meta: She considers not just one billiard table, but the universe of all possible billiard tables. And the kind of dynamics she studies doesn't directly concern the motion of the billiards on the table, but instead a transformation of the billiard table itself, which is changing its shape in a rule-governed way; if you like, the table itself moves like a strange planet around the universe of all possible tables . . . This isn't the kind of thing you do to win at pool, but it's the kind of thing you do to win a Fields Medal. And it's what you need to do in order to expose the dynamics at the heart of geometry; for there's no question that they're there.[13]

Mirzakhani gave an interview to *Quanta Magazine* in 2014 in which she reminisced on her youthful ambitions, comparing the process of researching mathematics to the process of writing a novel: "There are different characters, and you are getting to know them better," she said. "Things evolve, and then you look back at a character, and it's completely different from your first impression.[14]

In her brief lifetime, Mirzakhani won many honors in addition to the Fields Medal. She was an invited speaker on the topic of "Topology and Dynamical Systems & ODE" at the 2010 International Congress of Mathematicians in Hyderabad, India. Four years later she gave the lecture "On Weil-Petersson Volumes and Geometry of Random Hyperbolic Surfaces" at the 2014 International Congress of Mathematicians in Seoul, South Korea. She was elected to the Paris Academy of Sciences and the American Philosophical Society in 2015, the National Academy of Sciences in 2016, and the American Academy of Arts and Sciences in 2017.

Behind the scenes, however, Mirzakhani was struggling with her health. She was initially diagnosed with breast cancer in 2013; by 2016 the disease had spread to her liver and bones. Through it all, Mirzakhani continued working on mathematics and astounding her peers with her brilliance, creativity, tenacity, and humility. When she died in July 2017, it was a loss that reverberated around the world. According to her colleague, Ralph Cohen,

She not only was a brilliant and fearless researcher, but she was also a great teacher and terrific PhD adviser. Maryam embodied what being a mathematician or scientist is all about: the attempt to solve a problem that hadn't been solved before, or to understand something that hadn't been understood before. This is driven by a deep intellectual curiosity, and there is great joy and satisfaction with every bit of success. Maryam had one of the great intellects of our time, and she was a wonderful person.[15]

Iranian president Hassan Rouhani described her "unprecedented brilliance" as "a turning point in showing the great will of Iranian women and young people on the path towards reaching the peaks of glory."[16]

آرمان امروز

نمایندگان آمریکایی به گفت‌وگو با ایران رأی دادند

اقدام علیه ایران ممنوع

کرنش بی‌حساب جهان در برابر نابغه ریاضی

هدف: زدن شورا حذف امید

پای ترامپ روی طناب روی ردپان روس‌ها

شمارش معکوس برای کاهش سود بانکی

نه، جدا نمی‌شویم

مالیخولیای قدرت‌طلبی در مدیریت شهری

وام می‌گیرند تا عمل زیبایی کنند؟

برجام؛ نویدبخش موفقیت گفت‌وگو در حل پیچیده‌ترین مسائل هسته‌ای است

بیش از ۸ هزار کانال تلگرامی تصمیم کارگروه فیلترینگ نیست

اندوه ایران، حسرت جهان

مریم میرزاخانی، ملکه ریاضی دنیا درگذشت

داد نفتی

کودکان کار

تداوم ائتلاف انتخاباتی هم استراتژی، هم تاکتیک

اداره صداوسیما «هیأت امنایی» می‌شود

سرنخ‌های دیده نشده در قتل آتنا

جوان

۸ هزار کانال تلگرامی مجرمانه را ره

خرم‌آبادی: انسداد ۸ هزار کانال تلگرامی بیش از ۳۰ بار به‌موازات...

جامعه علمی جهان در سوگ نابغه ایرانی ریاضیات

ریشه‌های خشونت در جامعه

جوان‌گرایی در سپهر دانش

فرصت‌سازی از فیلم ساجری و تدبیر روز «عالی» بود

به کدامین ظریف معتقدید؟؟

امضای کری تضمین است

شهروندی

مذاکره بر سر آب با همسایگان

دریای مسجدالاقصی بسته است

ریشه‌های خشونت در جامعه

مریم میرزاخانی دنیای علم و سوگ کرد می‌رود

تسلیم جبر مرگ

تردد خریداران خرد در بازار مسکن

مقصر «ماجرای آتنا» کیست؟!

زندگی کودکان کار در مدرسه صبح رویش جوانه می‌زند

شهروند

روزنامه

مرگ آرام یک ذهن زیبا

دیدار اصلاح‌طلبان با رئیس جمهوری

اعتماد

ETEMAD

سیاست سعودی از نفت سعودی جداست

روز ناامنی اجتماعی و راه‌های برون رفت از بحران

رئیس سازمان بازرسی کل کشور خبر داد: اخراج بازنشستگان از ادارات دولتی

گفت‌وگو با قاتل آتنا ماهمدستت او نبودیم

انتقاضه جوانان جباران

حمله به یک روحانی

معادله زندگی «مریم»

مریم میرزاخانی (+) = مرگ (÷) = زن ایرانی (×) = ریاضی = نبوغ

CHELSEA WALTON

b. July 11, 1983

Searching for "New Truths"

"Mathematics is the alphabet with which God has written the universe."[1]

—Donald in Mathmagic Land, 1959

Left: An illustraton of Alice and the red and white chess queens from a 1921 songbook for Through the Looking-Glass *by author and mathematcian Charles Dodgson (aka Lewis Carroll). Themes from this story abound in Disney's math educaton featurette.*

Chelsea Walton grew up loving numbers. She also enjoyed puzzles and tinkering with patterns and symmetries. This passion was particularly evident when she created a frequency table of letters after counting all the As, Bs, Cs, and so on in a children's dictionary that her mother had given her.

Raised in Detroit, Michigan, she remembers watching *Donald in Mathmagic Land*, a twenty-seven-minute educational film from 1959 that was nominated for an Academy Award and ignited a passion for mathematics in generations of young people, including Walton. As she recalls, far from seeing math as dreaded busywork, she tackled arithmetic as if it were a game, and in high school she used a new invention—the Internet—to research careers in mathematics:

Above: The skyline of Detroit, Walton's hometown.

> *I remember waiting for that AOL dial-up modem to connect while I made my list of items to look up: "Mathematics + career," "Math + beautiful," and "Can I do logic puzzles all day and get paid for this?"* [2]

When she learned she could make a career of studying pure mathematics, Walton's path was set. She emailed math professors to learn more about what they did and how they got there, and she came to the conclusion that she would need a PhD.

Walton began her journey in pure mathematics by attending Michigan State University, where she met an instructor who would become her mentor. Professor Jeanne Wald studied a field of mathematics called noncommutative ring theory, which Walton went on to study after receiving a Bachelor of Science in mathematics with high honors. In 2007 she attended graduate school at the University of Michigan—Ann Arbor, taking the opportunity to complete research work in the United Kingdom at the University of Manchester with her thesis advisor, Toby Stafford, after her second year. She was also advised by ring theory expert Karen E. Smith (see page 144), producing the doctoral dissertation "On Degenerations and Deformations of Sklyanin Algebras" in 2011 and gaining her PhD in mathematics.

A MATHMAGICAL WONDERLAND

Donald Duck's strange journey through an alternate universe, where the trees have square roots and the ancient Greek mathematician Pythagoras holds jam sessions on the harp, was one of the most widely viewed educational films of the 1960s. The ornery cartoon duck enters the surreal landscape with shotgun in hand, suspicious of those math-loving "eggheads" and baffled by the numerical streams and geometric talking birds. However, when he discovers the connection between mathematics and music, and when he notices the mystical pentagram and golden rectangle in ancient monuments as well as petunias and waxflowers, Donald can't help but gape at the ubiquitous beauty and wonder of mathematics. After learning the calculations involved in a "diamond system" in billiards, he contemplates all the inventions—wheel, magnifying glass, spring, propeller—that would not have been possible without knowledge

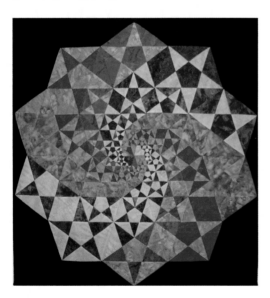

of basic geometry. Finally, Donald is alerted to another unique property of pentagrams: they can be drawn inside one another indefinitely and, thus, can generate infinitely complex fractal shapes. By the end of the featurette, he realizes that math isn't just for eggheads.

Left: This colorful quilt displays a repeating pentagram pattern alluded to in the classic 1959 film.

From 2011 to 2015, Walton had postdoctoral appointments at several prestigious institutions, including the University of Washington in Seattle, Washington, the Mathematical Sciences Research Institute in Berkeley, California, and the Massachusetts Institute of Technology in Cambridge, Massachusetts. In 2015, intrigued by the vibrant culture that reminded her of her hometown of Detroit, she moved to Philadelphia to become the Selma Lee Bloch Brown Professor of Mathematics at Temple University.

Walton has made a name for herself in work on algebraic structures, and, in collaboration with Susan Sierra of the University of Edinburgh, proved a conjecture in the theory of rings that had been unresolved for over twenty years. She has also received numerous grants from the National Science Foundation—high recognition for her work in topics including quantum symmetry, representation theory, Hopf algebras, and Nakayama automorphisms. The Alfred P. Sloan Foundation took note of her accomplishments and named her a 2017 Research Fellow, a highly competitive and prestigious honor identifying rising mathematicians and other scientists who have made significant marks on their field and represent the next generation of leaders in the US and Canada. Walton is just the fourth African American to be awarded a Sloan Fellowship in mathematics since the first fellowships were given in 1955.

Academia was exactly where Walton wanted to end up. In a 2017 profile she wrote for *Mathematically Gifted and Black*, published by the Network of Minorities in Mathematical Sciences, she reflected on her career path: "What drove me to academia (instead of industry) was the type of life I wanted. . . . I really like the lifestyle—math research is a creative job and the hours are not routine."[3] Walton also enjoys having a community of peers to whom she reports, rather than a direct supervisor. Most importantly, in academia, she is exposed to brilliant people from all over the world, representing a myriad of cultures and backgrounds. As she describes it:

Genuinely connecting with a wide variety of people (whether it's writing a paper together, having lunch, mentoring, being mentored,. . .), while being 100 percent myself, is one of my proudest achievements in math.[4]

WHAT IS
NONCOMMUTATIVE ALGEBRA?

One expects multiplication to produce the same result regardless of order. In Walton's field of noncommutative algebra, however, multiplication behaves in unusual ways. "That is, $a \times b$ doesn't have to be $b \times a$," she explained when she was awarded the 2017 Sloan Research Fellowship. "This occurs more often than one might think because functions are naturally noncommutative." To illustrate her point, she uses to the tedious task of laundry:

Wash your clothes first and then dry them; you'll get a different result if you dry your clothes first and then wash them. Or consider the order that you put on pieces of clothes—performing this in a different order produces very different appearances![5]

Another real-life example of a noncommutative action is trying to solve a Rubik's Cube, since the order in which you rotate the blocks is essential to achieving the solution. In mathematics, some operations are naturally noncommutative; in subtraction, for example, one minus two yields a different result than two minus one. Walton's area of expertise is in multiplication, where this is an abnormal phenomenon.[6]

Still early in her career, Walton is focused on learning new math, creating new math, and pushing forward. She plans to begin a tenured position at the University of Illinois Urbana-Champaign, where she will continue to study noncommutative mathematics and train a new generation of mathematicians. She is dedicated to increasing the representation of minorities in the mathematical sciences and teaches in programs designed to interest more young women to pursue mathematical research. At just thirty-four years old, Walton—who lives in central Illinois with her husband and two dogs—has sage advice for bright, ambitious young women and people of color:

The only person you should compare yourself with is "you yesterday." Keep pushing to be better and keep redefining "better" in the context of what makes you happy![7]

Below: The historic Altgeld Hall at the University of Illinois at Urbana-Champaign houses the department of mathematcs and its library.

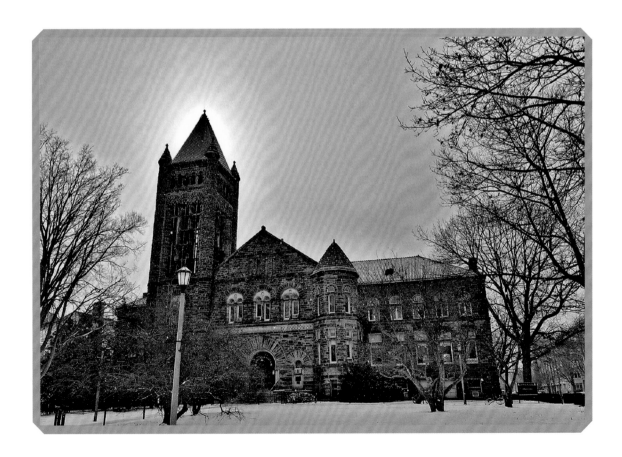

PAMELA E. HARRIS

b. November 28, 1983

Improving Diversity in the Mathematical Sciences

"Mathematics has taught me to be patient, to work hard, to be resilient, and to not take myself so seriously in this process."[1]

—Pamela E. Harris, 2017

Between first grade and twelfth grade, Pamela Harris had already attended close to ten different schools in two different countries. She spent her early childhood in Mexico and emigrated to California when she was eight years old, but things got tough financially. Her family briefly returned to Mexico before finally emigrating to Wisconsin when she was twelve years old. Recognizing her mathematical talent at a young age, Harris's father encouraged her intellectual pursuits the best way he could.

"I remember him giving me harder and harder division problems as a child, probably in an attempt to simply keep me busy," she says, "but his stories about the infinite size of the universe fascinated me and kept my love for mathematics alive."[2]

The number of schools she attended meant that teachers didn't know her long enough to notice that she might have some ability in mathematics. With teaching as a career goal, she set up a meeting with her high-school counselors, but they were unprepared to help her navigate options for a college education given her status as an undocumented youth. In a 2017 interview, she described the experience:

I vividly remember the meeting my father and I had with the counselor who told us that since I spoke fluent Spanish and English I should just go to the local technical college and become a bilingual secretary. She said this would be the best I could do. My father was furious and told her that I would go to that school and that I would do well enough to then go elsewhere and continue my studies. But that even if I "just" became a bilingual secretary, I would be the best.[3]

Harris used her taxpayer identification number, applied, and was accepted to Milwaukee Area Technical College, where she rediscovered her love for mathematics, taking classes in algebra, trigonometry, and calculus 1, 2, and 3. She married her husband and was able to adjust her immigration status, which then meant she could transfer to Marquette University. There she pursued a mathematics degree at Marquette, and met two life-changing mentors: Dr. Rebecca Sanders and Sanders's husband, Dr. Bill Rolli, who, at the time, were recent PhDs hired by Marquette. As Harris recalls,

Dr. Sanders taught me real analysis as an independent study course and during one of our meetings, she changed my life with a simple comment. She said, "When you go to graduate school . . ." At that time, I did not know about graduate school or that it was even possible for me to get a PhD in mathematics.[4]

Below: Part of Harris' PhD thesis includes this Hasse diagram associated to the zero weight Weyl alternation diagram of the exceptional Lie algebra G₂.

Dr. Sanders's encouragement led her to apply to graduate school, and it forever changed the course of Harris's life. Pursuing her advanced mathematics degree was an isolating experience, however. She had never met a fellow Latina in the field or a mathematician who shared a similar heritage and background. Despite the solitude, she persisted in her mathematical studies, focusing on combinatorics, a field of mathematics concerned with counting different things, such as the number of combinations in which a given number of objects can be arranged (remember factorials?). She explains her research in this way:

> *Consider the following combinatorial problem: In how many ways can the positive integer n be written as a sum of positive integers (ignoring the order)? For example, five can be written as a sum of positive integers in the following seven ways:*
>
> *5; 4 + 1; 3 + 2; 3 + 1 + 1; 2 + 2 + 1; 2 + 1 + 1 + 1; and 1 + 1 + 1 + 1 + 1*
>
> *Although this process is simple, determining a formula for the partition function, which counts the number of integer partitions of n, eluded generations of mathematicians and was only recently solved by Ken Ono, Jan Bruinier, Amanda Folsom, and Zach Kent in 2011. Their formula relied on the new and surprising discovery that partitions are fractal in nature.*[5]

Now an assistant professor in the department of mathematics and statistics at Williams College, Harris researches vector partition functions and projects in graph theory, work that has been supported through awards from the National Science Foundation and the Center for Undergraduate Research in Mathematics. "The impact of my mathematical research can be observed through the new combinatorial formulas I have found for the value of Kostant's partition function, a special vector partition function involved in the representation theory of Lie algebras,"[6] she explains. Representation theory is a branch of mathematics that represents the elements of algebraic structures as linear transformations of vector spaces.

Harris also enjoys working with undergraduates on mathematical research—especially students with a calculus background. "Working to help them understand how, as a mathematician, we can take a problem and generalize it further to find new results, is one of the most rewarding aspects of my job," she says.

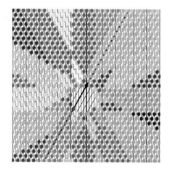

Above: The zero weight Weyl alternation diagram of the exceptional Lie algebra G_2 from Harris's PhD thesis.

Undergrads are so inquisitive and ask the right questions about why something is happening. It is also incredible to see such young minds make contributions to fields like representation theory by working through detailed examples, producing supporting evidence, which inform their conjectures, and much later, having learned needed background, be able to prove their conjectures.[7]

In 2012, Harris attended the National Conference of the Society for the Advancement of Chicanos/Hispanics and Native Americans in Science (SACNAS), and it changed her life. She is now part of a large, supportive community whose members uplift and help each other become leaders in their respective fields. "I have been very lucky to be surrounded by peers and mentors who support me professionally," she says.

[My peers] often remind me that as a Latina mathematician, my work has an impact outside of the walls of my institution and that I can make a difference in the mathematical community, both through my research and outreach activities. Their support has been invaluable throughout my career, and I am grateful to have them in my corner. I certainly wouldn't be where I am today without them.[8]

Below: Harris, with daughter Akira, believes a support system helped her achieve so many of her accomplishments.

Harris doesn't want anyone to feel the isolation she felt while doing her research and pursuing her degrees. She travels widely—her favorite perk of being a mathematician—to share research findings and to coorganize research symposia and professional development sessions for the national conference of SACNAS. She was an MAA Project NExT (New Experiences in Teaching) Fellow from 2012 to 2013 and is an editor of the e-Mentoring Network blog of the American Mathematical Society.

Harris's work has created new research opportunities for underrepresented students that support and reinforce their self-identity as scientists. In 2016, she helped develop and create the website www.Lathisms.org, an online platform that features prominently the extent of the research, teaching, and mentoring contributions of Latinos and Hispanics in the mathematical sciences.

NOTES & FURTHER READING

The information in this book comes from a mixture of original interviews, online sources, and books. For further research there are a number of good sources focused more solidly on mathematics and its history, including Agnes Scott College's "Biographies of Women Mathematicians" website (https://www.agnesscott.edu/lriddle/women/women.htm), the MacTutor History of Mathematics archive from the University of St. Andrews, Scotland (http://www-groups.dcs.st-and.ac.uk/~history/index.html), and Wolfram MathWorld (http://mathworld.wolfram.com/). For information on individual mathematicians and recent developments in the field, *Quanta Magazine* (https://www.quantamagazine.org/) is an excellent source. Additionally, NASA.gov is chock full of information and media related to women engineers and applied mathematicians.

GENERAL READING

Beery, Janet L., Sarah J. Greenwald, Jacqueline A. Jensen-Valin, and Maura B. Mast, eds. *Women in Mathematics: Celebrating the Centennial of the Mathematical Association of America.* Cham, Switzerland: Springer International Publishing, 2017.

Case, Bettye Anne and Anne M. Leggett, eds. *Complexities: Women in Mathematics.* Princeton, New Jersey: Princeton University Press, 2005.

Henrion, Claudia. *Women in Mathematics: The Addition of Difference.* Bloomington, Indiana: Indiana University Press, 1997.

Shetterly, Margot L. *Hidden Figures: The American Dream and the Untold Story of the Black Women Who Helped Win the Space Race.* New York: William Morrow, 2016.

PART I

---

1. Ann Phillips (ed.), *A Newnham Anthology* (Cambridge, U.K.: Cambridge University Press, 1979): 33–34.

WANG ZHENYI

1. Gabriella Bernardi, "Wang Zhenyi (1768–1797)" in *Female Astronomers and Scientists before Caroline Herschel* (Cham, Switzerland: Springer Praxis Books, 2016), https://doi.org/10.1007/978-3-319-26127-0_23.
2. Ibid.
3. "Wang Zhenyi," Lunar and Planetary Institute of the Universities Space Research Association, https://www.lpi.usra.edu/resources/vc/vcinfo/?refnum=262.
4. Bernardi, "Wang Zhenyi (1768–1797).
5. "Wang Zhenyi: Astronomer, Mathematician, and Poet," April Magazine, August 22, 2017, http://www.aprilmag.com/2017/08/22/wang-zhenyi-astronomer-mathematician-and-poet/.

SOPHIE GERMAIN

1. Nick Mackinnon, "Sophie Germain, or, Was Gauss a feminist?" *The Mathematical Gazette* 74, no. 470 (Dec. 1990): 346–351.
2. The term "mod *P*" refers to the modular arithmetic developed by Carl Friedrich Gauss in which integers wrap around after reaching a value known as the modulus (in this case, *P*). For

example, 12 is the modulus of our standard 12-hour clock, with the hours reverting to 1 p.m. or 1 a.m. after the short hand reaches the 12 at the top of the clock, representing noon as well as midnight.

3. Andrea Del Centina, "Unpublished manuscripts of Sophie Germain and a revaluation of her work on Fermat's Last Theorem," *Archive for History of Exact Sciences* 62, no. 4 (July 2008): 349–392.

4. British mathematician Sir Andrew Wiles is credited with proving Fermat's Last Theorem. The full proof in its final form was presented in a paper titled "Modular Elliptic Curves and Fermat's Last Theorem," published in the May 1995 issue of the journal *Annals of Mathematics*. For an abbreviated account of the entire quest and controversy, see Eric W. Weisstein's *Wolfram MathWorld* article titled "Fermat's Last Theorem": http://mathworld. wolfram.com/FermatsLastTheorem.html.

5. Arthur Engel, *Problem-Solving Strategies* (New York: Springer Verlag, 1998), 121.

For more information on Sophie Germain, consult the following:

Laubenbacher, Reinhard and David Pengelley. *Mathematical Expeditions: Chronicles by the Explorers*. New York: Springer-Verlag, 1999.

Laubenbacher, Reinhard and David Pengelley. "Voici ce que j'ai trouvé": Sophie Germain's grand plan to prove Fermat's Last Theorem." Preprint for *Historia Mathematica*, January 24, 2010. https://www.math.nmsu.edu/~davidp/germain06-ed.pdf

Laubenbacher, Reinhard and David Pengelley. "'Voici ce que j'ai trouvé': Sophie Germain's Grand Plan to Prove Fermat's Last Theorem." Teaching with Original Historical Sources in Mathematics. New Mexico State University. 2010. http://emmy.nmsu.edu/~history/germain. html

Noyce, Pendred. *Remarkable Minds*. Boston: Tumblehome Learning, 2016: 40–47.

O'Connor, J.J. and E.F. Robertson. "Marie-Sophie Germain." School of Mathematics and Statistics, University of St Andrews, Scotland. http://www-history.mcs.st-andrews.ac.uk/ Biographies/Germain.html.

Riddle, Larry. "Sophie Germain." Biographies of Women Mathematicians. Agnes Scott College. Last modified February 25, 2016. https://www.agnesscott.edu/lriddle/women/germain.htm

Singh, Simon. "Math's Hidden Women." NOVA, PBS, October 28, 1997. https://www.pbs.org/ wgbh/nova/physics/sophie-germain.html

SOFIA KOVALEVSKAYA

1. "Letter to Madame Schabelskoy" in *Sónya Kovalévsky: Her Recollections of Childhood*, trans. Isabel F. Hangood (New York: The Century Co., 1895).

2. Sofia went by the name "Sonya Kovalevsky" after her arrival in Stockholm in 1883.

3. "Letter to Madame Schabelskoy" in *Sónya Kovalévsky: Her Recollections of Childhood*, trans. Isabel F. Hangood (New York: The Century Co., 1895)

4. Ibid.

For more information on Sofia Kovalevskaya, consult the following:

Burslem, Tom. "Sofia Vasilevna Kovalevskaya." School of Mathematics and Statistics, University of St Andrews, Scotland. http://www-groups.dcs.st-andrews.ac.uk/history/Miscellaneous/Kovalevskaya/biog.html.

Koblitz, Ann Hibner. *A Convergence of Lives: Sofia Kovalevskaia: Scientist, Writer, Revolutionary.* New Brunswick, N.J.: Rutgers University Press, 1993.

Lindenberg, Katja. "Sofia." University of California, San Diego. http://hypatia.ucsd.edu/~kl/kovalevskaya.html.

"Mathematician Sofia Kovalevskaya - her life and legacy." h2g2 The Hitchhiker's Guide to the Galaxy: Earth Edition, August 16, 2009. Last modified September 23, 2009. https://h2g2.com/entry/A55840692.

"Mathematics opens up a new, wonderful world." Annette Vogt, interviewed by Beate Koch. Max-Planck-Gesellschaft. March 18, 2017. https://www.mpg.de/female-pioneers-of-science/sofia-kovalevskaya.

Meares, Kimberly A. "The Works of Sonya Kovalevskaya." http://www.pdmi.ras.ru/EIMI/2000/sofia/SKpaper.html.

Noyce, Pendred. *Magnificent Minds.* Boston: Tumblehome Learning, 2016: 64–71.

O'Connor, J.J. and E.F. Robertson. "Sofia Vasilyevna Kovalevskaya." School of Mathematics and Statistics, University of St Andrews, Scotland. http://www-history.mcs.st-and.ac.uk/Biographies/Kovalevskaya.html.

Pickover, Clifford A. *The Math Book.* New York: Sterling, 2009: 260.

Rappaport, Karen D. "S. Kovalevsky: A Mathematical Lesson." *The American Mathematical Monthly* 88 (October 1981): 564–573.

Riddle, Larry. "Sofia Kovalevskaya." Biographies of Women Mathematicians. Agnes Scott College. Last modified February 25, 2016. https://www.agnesscott.edu/lriddle/women/kova.htm.

WINIFRED EDGERTON MERRILL

1. Larry Riddle, "Winifred Edgerton Merrill," Biographies of Women Mathematicians, Agnes Scott College, last modified February 25, 2016, https://www.agnesscott.edu/lriddle/women/merrill.htm.

2. Bronwyn Knox, "She Opened The Door," The Low Down, Columbia University, last modified November 16, 2017, http://thelowdown.alumni.columbia.edu/she_opened_the_door.

3. Susan E. Kelly and Sarah A. Rozner, "Winifred Edgerton Merrill: 'She Opened the Door,'" *Notices of the AMS* 59, no. 4 (April 2014): 504–512, http://dx.doi.org/10.1090/noti818.

4. Ibid.

5. Ibid.

6. W.C. Winlock, "The Pons-Brooks Comet," *Science* 3, no. 50 (January 18, 1884), 67–69, http://science.sciencemag.org/content/ns-3/50/67.

7. Elizabeth Roemer, "Jean Louis Pons, Discoverer of Comets," *Astronomical Society of the Pacific Leaflets* 8, no. 371 (May 1960): 159–166, http://adsabs.harvard.edu/full/1960ASPL....8..159R.

8. Kelly, "Winifred Edgerton Merrill: 'She Opened the Door.'"

9. Ibid.

For more information on Winifred Edgerton Merrill, consult the following:

Sawyer Hogg, Helen. "Out of Old Books." Journal of the Royal *Astronomical Society of Canada* 48, no. 2, 74 (April 1954): 74–76. http://adsabs.harvard.edu/full/1954JRASC..48...74S.

EMMY NOETHER

1. Albert Einstein, "Professor Einstein Writes in Appreciation of a Fellow-Mathematician," *The New York Times* (May 5, 1935).
2. The study of algebraic structures, including groups, rings, fields, modules, vector spaces, and algebras (i.e., vector spaces containing a bilinear product).
3. August Dick, *Emmy Noether 1882–1935* (Boston: Birkhiiuser Boston, 1981).
4. Natalie Angier, "The Mighty Mathematician You've Never Heard Of," *The New York Times* (March 26, 2012).
5. Leon M. Lederman and Christopher T. Hill, *Symmetry and the Beautiful Universe* (Amherst, New York: Prometheus Books, 2004).
6. Ibid.
7. A class of algebraic structures in which the order of operations affects the results.
8. M.F. Atiyah and I.G. MacDonald, *Introduction to Commutative Algebra* (Reading, Massachusetts: Westview Press, 1969).
9. Peter Roquette, *Emmy Noether and Hermann Weyl* (January 28, 2008), https://www.mathi.uni-heidelberg.de/~roquette/weyl+noether.pdf.
10. Ibid.

For more information on Emmy Noether, consult the following:

Aczel, Amir. *A Strange Wilderness.* New York: Sterling 2012.

Brewer, James W and Marth K. Smith, eds. *Emmy Noether: A Tribute to Her Life and Work.* New York: Marcel Dekker, 1981.

"Emmy Noether Biography." TheFamousPeople.com. Last modified November 14, 2017. https://www.thefamouspeople.com/profiles/emmy-noether-507.php.

Noyce, Pendred. *Magnificent Minds.* Boston: Tumblehome Learning, 2016: 94–99.

O'Connor, J.J. and E.F. Robertson. "Emmy Amalie Noether." School of Mathematics and Statistics, University of St Andrews, Scotland. http://www-history.mcs.st-and.ac.uk/Biographies/Noether_Emmy.html.

Pickover, Clifford A. *The Math Book.* New York: Sterling, 2009: 342.

Taylor, Peter. "Emmy Noether (1882-1935)." Australian Mathematics Trust. March 1999. http://www.amt.edu.au/biognoether.html.

EUPHEMIA HAYNES

1. Euphemia Haynes, "M St. School Valedictorian Speech," 1907 (Box 9, Folder 7), Haynes-Lofton Family Papers, The Catholic University of America, Washington, D.C.
2. Euphemia Haynes, "Significance of Mathematics in the War and the Post-War World," *The Journal of the College Alumnae Club*, October 1943 (Box 40, Folder 1), Haynes Lofton Family Papers, The Catholic University of America, Washington, D.C.

For more information on Euphemia Haynes, consult the following:

Kelly, Susan E., Carly Shinners, and Katherine Zoroufy. "Euphemia Lofton Haynes: Bringing Education Closer to the 'Goal of Perfection.'" March 2, 2017. https://arxiv.org/pdf/1703.00944.pdf.

Kelly Susan E., Carly Shinners, and Katherine Zoroufy. "Euphemia Lofton Haynes: Bringing Education Closer to the 'Goal of Perfection.'" *Notices of the AMS 64,* no. 9 (October 2017). http://dx.doi.org/10.1090/noti1579.

Pitts, Vanessa. "Haynes, Martha Euphemia Lofton (1890-1980)." BlackPast.org. http://www.blackpast.org/aah/haynes-martha-euphemia-lofton-1890-1980.

Williams, Scott W. "Martha Euphemia Lofton Haynes." Black Women in Mathematics. Mathematics Department, State University of New York, Buffalo. Last modified July 1, 2001. http://www.math.buffalo.edu/mad/PEEPS/haynes.euphemia.lofton.html.

PART II

GRACE HOPPER

1. Reported in the September 29, 1967, issue of Vassar College's *Miscellany News*, when Grace Hopper delivered the keynote speech called "Computers and Your Future," at the dedication of a new computer center and its IBM 360 computer at the school: https://chronology.vassar.edu/records/1967/1967-09-29-ibm-computers.html.

2. "USS Hopper," Pacific Fleet Surface Ships, U.S. Navy, http://www.public.navy.mil/surfor/ddg70/Pages/namesake.aspx.

3. Kathleen Williams, "Improbable Warriors: Mathematicians Grace Hopper and Mina Rees in World War II," in *Mathematics and War*, eds. Bernhelm Booß-Bavnbek and Jens Høyrup (Basel: Springer Birkhäuser, 2003): 117, https://doi.org/10.1007/978-3-0348-8093-0_5.

4. "Grace Murray Hopper," Grace Hopper Celebration of Women in Computing 1994 Conference Proceedings, http://www.cs.yale.edu/homes/tap/Files/hopper-story.html.

For more information on Grace Hopper, consult the following:

"IBM's ASCC Introduction." IBM. https://www-03.ibm.com/ibm/history/exhibits/markI/markI_intro.html.

Norman, Rebecca. "Grace Murray Hopper." Biographies of Women Mathematicians. Agnes Scott College. Last modified February 25, 2016. https://www.agnesscott.edu/lriddle/women/hopper.htm.

O'Connor, J.J. and E.F. Robertson. "Grace Brewster Murray Hopper." School of Mathematics and Statistics, University of St Andrews, Scotland. http://www-history.mcs.st-andrews.ac.uk/Biographies/Hopper.html.

Office of Public Affairs & Communications. "Grace Murray Hopper (1906–1992): A legacy of innovation and service." Yale News, Yale University. February 10, 2017. https://news.yale.edu/2017/02/10/grace-murray-hopper-1906-1992-legacy-innovation-and-service.

Vassar Historian. "Grace Murray Hopper." Vassar Encyclopedia, Vassar College. Last modified 2013. http://vcencyclopedia.vassar.edu/alumni/grace-murray-hopper.html.

Vassar Historian. "September 29, 1967." A Documentary Chronicle of Vassar College, Vassar
College. https://chronology.vassar.edu/records/1967/1967-09-29-ibm-computers.html.

MARY GOLDA ROSS

1. "The Cherokee Nation Remembers Mary Golda Ross, the First Woman Engineer for
   Lockheed," Cherokee Nation News Release, Cherokee Nation Director of Communications
   (May 13, 2008), http://www.cherokee.org/News/Stories/23649.
2. Laurel M. Sheppard, "An Interview with Mary Ross," Portfolio: Profiles & Biographies, Lash
   Publications International, http://www.nn.net/lash/maryross.htm.
3. Ibid.
4. "Skunk Works® Origin Story," Lockheed Martin, https://www.lockheedmartin.com/us/
   aeronautics/skunkworks/origin.html.
5. Sandberg, Ariel. "Remembering Mary Golda Ross." The Michigan Engineer News Center,
   June 4, 2017. https://news.engin.umich.edu/2017/06/remembering-mary-golda-ross/. (was
   originally Sheppard, "An Interview with Mary Ross")

For more information on Mary Golda Ross, consult the following:

Blakemore, Erin. "This Little-Known Math Genius Helped America Reach the Stars."
Smithsonian.com, March 29, 2017. http://www.smithsonianmag.com/smithsonian-institution/
little-known-math-genius-helped-america-reach-stars-180962700/.

Briggs, Kara. "Mary G. Ross blazed a trail in the sky as a woman engineer in the space race,
celebrated museum." National Museum of the American Indian, October 7, 2009. http://blog.
nmai.si.edu/main/2009/10/mary-g-ross-blazed-a-trail-in-the-sky-as-a-woman-engineer-in-
the-space-race-celebrated-museum-.html.

"Feature Detail Report for: Park Hill." Geographic Names Information System. USGS. https://
geonames.usgs.gov/apex/f?p=gnispq:3:0::NO::P3_FID:1096432.

"Learning with the Times: Tough to intercept ballistic missiles." The Times of India, December
6, 2010. https://timesofindia.indiatimes.com/india/Learning-with-the-Times-Tough-to-
intercept-ballistic-missiles/articleshow/7050349.cms.

"Mary G. Ross." San Jose Mercury News, May 6, 2008. http://www.legacy.com/obituaries/
mercurynews/obituary.aspx?pid=109118876.

"Mary G. Ross." YourDictionary. http://biography.yourdictionary.com/mary-g-ross.

"Ms. Mary G. Ross." Hall of Fame Members. Silicon Valley Engineering Council. http://svec.
herokuapp.com/hall-of-fame.

Ouimette-Kinney, Mary. "Mary G. Ross." Biographies of Women Mathematicians. Agnes Scott
College. Last modified July 10, 2016. https://www.agnesscott.edu/lriddle/women/maryross.htm.

Sandberg, Ariel. "Remembering Mary Golda Ross." The Michigan Engineer News Center, June
14, 2017. https://news.engin.umich.edu/2017/06/remembering-mary-golda-ross/.

Scott, Jeff. "Bombs, Rockets & Missiles." Aerospaceweb.org, January 16, 2005. http://www.
aerospaceweb.org/question/weapons/q0211.shtml.

Toomey, David F. "Can This P-38 Be Saved?" *Air & Space Magazine*, November 2009. https://www.airspacemag.com/history-of-flight/can-this-p-38-be-saved-137693818/

What's My Line? "*What's My Line?* - Andy Griffith; Jack Lemmon [panel] (Jun 22, 1958)." YouTube, February 8, 2014. https://www.youtube.com/watch?v=vFlvpMf-dIo.

Williams, Jasmin K. "Mary Golda Ross: The first Native American female engineer." *New York Amsterdam News*, March 21, 2013. http://amsterdamnews.com/news/2013/mar/21/mary-golda-ross-the-first-native-american-female/.

DOROTHY VAUGHAN

1. Carson Reeher, "From Moton to NASA," *The Farmville Herald*, January 12, 2017, http://www.farmvilleherald.com/2017/01/from-moton-to-nasa/.

2. Margot Shetterly, *Hidden Figures: The American Dream and the Untold Story of the Black Women Mathematicians Who Helped Win the Space Race* (New York: William Morrow, 2016).

3. Ibid.

4. Beverly E. Golemba, *Human Computers: The Women in Aeronautical Research* (Unpublished manuscript: NASA Langley Archives, 1994), https://crgis.ndc.nasa.gov/crgis/images/c/c7/Golemba.pdf.

For more information on Dorothy Vaughan, consult the following:

"Dorothy Johnson Vaughan." Biography.com. Last modified November 14, 2016. https://www.biography.com/people/dorothy-johnson-vaughan-111416.

"Dorothy Vaughan." The Gallery of Heroes. http://thegalleryofheroes.com/topic/Dorothy-Vaughan.

"Scout Launch Vehicle Program." Langley Research Center, National Aeronautics and Space Administration. https://www.nasa.gov/centers/langley/news/factsheets/Scout.html.

Shetterly, Margot Lee. "Katherine Johnson Biography." From Hidden to Modern Figures, National Aeronautics and Space Administration. Last modified August 3, 2017. https://www.nasa.gov/content/dorothy-vaughan-biography.

KATHERINE JOHNSON

1. Margot Lee Shetterly, "Katherine Johnson Biography," From Hidden to Modern Figures, National Aeronautics and Space Administration, last modified August 3, 2017, https://www.nasa.gov/content/katherine-johnson-biography.

2. Margot Shetterly, *Hidden Figures: The American Dream and the Untold Story of the Black Women Mathematicians Who Helped Win the Space Race* (New York: William Morrow, 2016).

3. Heather S. Deiss/NASA Educational Technology Services, "Katherine Johnson: A Lifetime of STEM," NASA Langley, National Aeronautics and Space Administration, November 6, 2013, last modified August 7, 2017, https://www.nasa.gov/audience/foreducators/a-lifetime-of-stem.html.

4. Wini Warren, *Black Women Scientists in the United States* (Bloomington, Indiana: Indiana University Press, 2000).

5. Mark Bloom, "Apollo 13 Astronauts Return After Deep Space Crisis in 1970," *New York Daily*

News, April 18, 1970, http://www.nydailynews.com/news/national/apollo-13-astronauts-return-deep-space-crisis-1970-article-1.2177667.

6. C-SPAN, "Katherine Johnson receives the Medal of Freedom," November 24, 2015.

7. Charles Bolden, "Katherine Johnson, the NASA Mathematician Who Advanced Human Rights with a Slide Rule and Pencil," *Vanity Fair*, September 2016, https://www.vanityfair.com/culture/2016/08/katherine-johnson-the-nasa-mathematician-who-advanced-human-rights.

For more information on Katherine Johnson, consult the following:

"Apollo 13." Apollo 13, National Aeronautics and Space Administration, July 8, 2009. Last modified August 7, 2017. https://www.nasa.gov/mission_pages/apollo/missions/apollo13.html.

"Apollo 13 (AS-508)." The Apollo Program, Smithsonian National Air and Space Museum. https://airandspace.si.edu/explore-and-learn/topics/apollo/apollo-program/landing-missions/apollo13.cfm.

Bartels, Meghan. "The unbelievable life of the forgotten genius who turned Americans' space dreams into reality." *Business Insider*, August 22, 2016. http://www.businessinsider.com/katherine-johnson-hidden-figures-nasa-human-computers-2016-8.

Cellania, Miss. "Forty Years Ago: Apollo 13." Mental Floss, April 13, 2010. http://mentalfloss.com/article/24441/forty-years-ago-apollo-13.

Gambino, Lauren. "NASA facility honors African American woman who plotted key space missions." *The Guardian*, September 22, 2017. https://www.theguardian.com/science/2017/sep/22/hidden-figures-mathematician-katherine-johnson-nasa-facility-open.

Lewin, Sarah. "NASA Langley's Katherine G. Johnson Computational Research Facility Opens." Space.com, September 25, 2017. https://www.space.com/38261-nasa-katherine-johnson-computational-facility-opens.html.

"Thomas O. Paine." National Aeronautics and Space Administration. Last modified October 22, 2004. https://history.nasa.gov/Biographies/paine.html.

White House, The: Office of the Press Secretary. "Remarks by the President at the Congressional Black Caucus 45th Annual Phoenix Awards Dinner." September 20, 2015. https://obamawhitehouse.archives.gov/the-press-office/2015/09/21/remarks-president-congressional-black-caucus-45th-annual-phoenix-awards.

MARY WINSTON JACKSON

1. Richard Stradling, "Retired Engineer Remembers Segregated Langley," Daily Press, February 08, 1998, http://articles.dailypress.com/1998-02-08/news/9802080064_1_hampton-institute-aeronautical-engineer-cafeteria.

2. Ibid.

3. "Traditions," Girl Scouts of the United States of America, http://www.girlscouts.org/en/about-girl-scouts/traditions.html.

4. "History," Hampton University, http://www.hamptonu.edu/about/history.cfm.

5. Margot Lee Shetterly, "Mary Jackson Biography," From Hidden to Modern Figures, National Aeronautics and Space Administration, last modified August 3, 2017, https://www.nasa.gov/content/mary-jackson-biography.

6. "Mary W. Jackson Federal Women's Program Coordinator," Internal Memo, Langley Research Center, National Aeronautics and Space Administration, October 1979, https://crgis.ndc.nasa.gov/crgis/images/9/96/MaryJackson1.pdf.

For more information on Mary Winston Jackson, consult the following:

Czarnecki, K. R. and Mary W. Jackson. *Effects of Nose Angle and Mach Number on Transition on Cones at Supersonic Speeds.* National Advisory Committee for Aeronautics, Technical Note 4388. https://ntrs.nasa.gov/archive/nasa/casi.ntrs.nasa.gov/19930085290.pdf.

"George P. Phenix High School Story, The." George P. Phenix High School. https://www.phenixhighstory.org/school-background/.

"Isentropic Flow." Glenn Research Center, National Aeronautics and Space Administration. https://www.grc.nasa.gov/www/k-12/airplane/isentrop.html.

"Mary Jackson." The Gallery of Heroes. http://thegalleryofheroes.com/topic/Mary-Jackson.

"Mary Winston Jackson." Daily Press via Legacy.com, February 16, 2005. http://www.legacy.com/obituaries/dailypress/obituary.aspx?pid=3163015.

Momodu, Samuel. "Jackson, Mary Winston (1921–2005)." Blackpast.org. http://www.blackpast.org/aah/jackson-mary-winston-1921-2005.

SHAKUNTALA DEVI

1. "Shakuntala Devi," Obituaries, *The Telegraph*, April 22, 2013, http://www.telegraph.co.uk/news/obituaries/10011281/Shakuntala-Devi.html.
2. Haresh Pandya, "Shakuntala Devi, 'Human Computer' Who Bested the Machines, Dies at 83," *The New York Times*, April 23, 2013, http://www.nytimes.com/2013/04/24/world/asia/shakuntala-devi-human-computer-dies-in-india-at-83.html.
3. Chloe Albanesius, "Shakuntala Devi, the 'Human Computer,' gets Google Doodle, " *PC Magazine*, November 4, 2013
4. "Shakuntala Devi," Obituaries, *The Telegraph*.

For more information on Shakuntala Devi, consult the following:

Eveleth, Rose. "Math Prodigy Shakuntala Devi, 'The Human Computer,' Dies at 83." Smithsonian.com, April 23, 2013. https://www.smithsonianmag.com/smart-news/math-prodigy-shakuntala-devi-the-human-computer-dies-at-83-38972900/.

Jones, Maggie. "Shakuntala Devi." The Lives They Lived, *The New York Times Magazine*. https://www.nytimes.com/news/the-lives-they-lived/2013/12/21/shakuntala-devi/

MaxMediaAsia. "India's Human Computer Shakuntala Devi." YouTube, March 24, 2013. https://www.youtube.com/watch?v=l0fXPzcbmPk.

Tiwari, Shewali. " 'Here's Everything You Need to Know About the 'Human Computer,' Shakuntala Devi." *India Times*, April 21, 2017. https://www.indiatimes.com/news/india/here-s-everything-you-need-to-know-about-the-human-computer-shakuntala-devi-276144.html.

"World's Fastest 'Human Computer' Maths Genius Dies, Aged 83 in India." *The Daily Mail*, April 22, 2013. http://www.dailymail.co.uk/news/article-2312987/Human-dies-Shakuntala-Devi-Indian-maths-genius-dead-83.html.

ANNIE EASLEY

1. "Annie J. Easley, interviewed by Sandra Johnson," NASA Headquarters Oral History Project: Edited Oral History Transcript, August 21, 2001, last modified July 16, 2010, https://www.jsc.nasa.gov/history/oral_histories/NASA_HQ/Herstory/EasleyAJ/EasleyAJ_8-21-01.htm.
2. Ibid.
3. Ibid.
4. Ibid.

For more information on Annie Easley, consult the following:

Lee, Nicole. "Annie Easley helped make modern spaceflight possible." Engadget.com. February 13, 2015. https://www.engadget.com/2015/02/13/annie-easley/.

Mills, Anne K. "Annie Easley, Computer Scientist." National Aeronautics and Space Administration. September 21, 2015, last modified August 7, 2017. https://www.nasa.gov/feature/annie-easley-computer-scientist.

MARGARET HAMILTON

1. A.J.S. Rayl, "NASA Engineers and Scientists-Transforming Dreams Into Reality," *50th Magazine, National Aeronautics and Space Administration*, last modified October 16, 2008, https://www.nasa.gov/50th/50th_magazine/scientists.html.
2. Ibid.

For more information on Margaret Hamilton, consult the following:

Kingma, Luke and Jolene Creighton. "Margaret Hamilton: The Untold Story of the Woman Who Took Us to the Moon." Futurism. July 20, 2016. Last modified November 19, 2016. https://futurism.com/margaret-hamilton-the-untold-story-of-the-woman-who-took-us-to-the-moon/.

Russo, Nicholas P. "Margaret Hamilton, Apollo Software Engineer, Awarded Presidential Medal of Freedom." NASA History, National Aeronautics and Space Administration. November 22, 2016. Last modified August 6, 2017. https://www.nasa.gov/feature/margaret-hamilton-apollo-software-engineer-awarded-presidential-medal-of-freedom.

PART III

1. Hillary Fennell, "This much I know: Aoibhinn Ní Shúilleabháin," *Irish Examiner*, November 5, 2016, http://www.irishexaminer.com/lifestyle/features/this-much-i-know-aoibhinn-ni-shuilleabhain-428996.html.
2. Helena Kaschel, "Don't call me a prodigy: the rising stars of European mathematics," *Deutsche Welle* dw.com, July 23, 2016, http://www.dw.com/en/dont-call-me-a-prodigy-the-rising-stars-of-european-mathematics/a-19421389.
3. 7th European Congress of Mathematics, "Laureates," July 18–22, 2016, http://euro-math-soc.eu/system/files/news/Druck%20-%20pp_3657_Broschuere%20-%207ECM.pdf.
4. Kenschaft, Patricia Clark, *Change is Possible: Stories of Women and Minorities in Mathematics* (Providence, Rhode Island: American Mathematical Society, 2005).

5. Mary Beth Ruskai, "Gray Receives AAAS Mentor Award," *Notices of the American Mathematical Society* 42, no. 4 (April 1995): 466.

6. "Lectures," Association for Women in Mathematics, https://sites.google.com/site/awmmath/ programs/lectures.

SYLVIA BOZEMAN

1. Abigail Meisel, "Math Master: Sylvia T. Bozeman, MA'70, Honored with National Medal of Science Committee Appointment," *Vanderbilt Magazine*, Vanderbilt University, November 20, 2016, https://news.vanderbilt.edu/vanderbiltmagazine/math-master-sylvia-t-bozeman-ma70-honored-with-national-medal-of-science-committee-appointment/.

2. "Sylvia Bozeman," Mathematically Gifted and Black, February 7, 2017, http://www.mathematicallygiftedandblack.com/profiles/February_7.html.

3. Peggy Mihelich, "Women in mathematics: Professor Sylvia Bozeman," American Association for the Advancement of Science, December 13, 2010, last modified August 1, 2016, https://www.aaas.org/blog/member-spotlight/women-mathematics-professor-sylvia-bozeman.

4. "President Obama Appoints Professor Emerita to Key Administration Post," Our Stories, Spelman College, July 2016, https://www.spelman.edu/about-us/news-and-events/our-stories/stories/2016/07/11/president-obama-appoints-professor-emerita-to-key-administration-post.

5. "Sylvia Bozeman," Mathematically Gifted and Black.

6.. Meisel, "Math Master."

7. "Sylvia Bozeman," Mathematically Gifted and Black.

8. Mihelich, "Women in mathematics."

9. "Sylvia Bozeman," Mathematically Gifted and Black.

For more information on Sylvia Bozeman, consult the following:

Bozeman, Sylvia, Rhonda Hughes, and Ami Radunskaya. "The EDGE Program: Adding Value through Diversity." http://www.brynmawr.edu/math/people/rhughes/9.pdf.

Houston, Johnny L. "Sylvia Trimble Bozeman." Mathematical Association of America. 1997. https://www.maa.org/programs/underrepresented-groups/summa/summa-archival-record/sylvia-trimble-bozeman.

Morrow, Charlene and Teri Peri. *Notable Women in Mathematics: A Biographical Dictionary.* Westport, Connecticut: Greenwood Press, 1998.

"Sylvia Bozeman." ScienceMakers, The HistoryMakers: *The Nation's Largest African American Oral History Collection.* December 18, 2012. http://www.thehistorymakers.org/biography/sylvia-bozeman.

FERN Y. HUNT

1. Interview, October 2017.

2. "Fern Y. Hunt," Mathematics Quiz, Kids' Zone, Learning with NCES. https://nces.ed.gov/nceskids/grabbag/Mathquiz/mathresult.asp?coolest=j.

3. Interview, October 2017.

4. Ibid.

5. Ibid.

For more information on Fern Hunt, consult the following:

Ambruso, Kathleen. "Fern Y. Hunt." Mathematical Association of America. https://www.maa.org/fern-y-hunt.

"Fern Hunt." Mathematical Research and the World of Nature. Association for Women in Mathematics. http://www.awm-math.org/ctcbrochure/hunt.html.

"Fern Hunt." ScienceMakers, *The HistoryMakers: The Nation's Largest African American Oral History Collection*. September 14, 2012. http://www.thehistorymakers.org/biography/fern-hunt.

Richardson, Alicia. "Dr. Fern Hunt: Mastering Chaos in Theory and in Life." 2002 Association for Women in Mathematics Essay Contest. Last modified July 8, 2010. https://sites.google.com/site/awmmath/programs/essay-contest/contest-rules/essay-contest-past-results/essays/drfernhuntmasteringchaosintheoryandinlife.

Williams, Scott W. "Fern Y. Hunt." Black Women in Mathematics. Mathematics Department, State University of New York, Buffalo. http://www.math.buffalo.edu/mad/PEEPS/hunt_ferny.html.

MARIA KLAWE

1 Lucas Laursen, "No, You're Not an Impostor," Science, February 15, 2008, http://www.sciencemag.org/careers/2008/02/no-youre-not-impostor.

2. Interview, October 2017.

3. Maria Klawe, "Impostoritis: A Lifelong, but Treatable, Condition," *Slate*, March 24, 2014, http://www.slate.com/articles/technology/future_tense/2014/03/imposter_syndrome_how_the_president_of_harvey_mudd_college_copes.html.

4. Maria Klawe, "A Fifty Year Wave of Change," Women in Technology, O'Reilly, September 5, 2007, http://archive.oreilly.com/pub/a/womenintech/2007/09/05/a-fifty-year-wave-of-change.html.

For more information on Maria Klawe, consult the following:

"Biography of President Maria Klawe." Harvey Mudd College. https://www.hmc.edu/about-hmc/president-klawe/biography-of-president-maria-klawe/.

Edwards, Martha. "16 Celebrity Quotes on Suffering with Imposter Syndrome." *Marie Claire*, November 11, 2016. http://www.marieclaire.co.uk/entertainment/celebrity-quotes-on-impostor-syndrome-434739.

"Interview with Maria Klawe." Computing Research Association—Women. October 20, 2012. https://cra.org/cra-w/interview-with-maria-klawe/.

"Maria's Art." Harvey Mudd College. https://www.hmc.edu/about-hmc/president-klawe/biography-of-president-maria-klawe/marias-art/.

AMI RADUNSKAYA

1. Radunskaya, Ami, "Mathematical Biology: A Personal Journey," *SMB Newsletter* 28, no. 3 (September 2015): 9–10, http://www.smb.org/publications/newsletter/bios/vol28n03_radunskaya.pdf.
2. "Ami Radunskaya, Pomona College," 2015 Elections, Association for Women in Mathematics, https://sites.google.com/site/awmmath/home/awm-elections-2015/president-elect.
3. Ginger Pinholster, "Mathematician Ami Radunskaya Wins 2016 AAAS Mentor Award," American Association for the Advancement of Science, November 29, 2016, https://www.aaas.org/news/mathematician-ami-radunskaya-wins-2016-aaas-mentor-award.

For more information on Ami Radunskaya, consult the following:
"Ami E. Radunskaya." Directory, Pomona College. https://www.pomona.edu/directory/people/ami-e-radunskaya.
IndieFlix. The Empowerment Project. http://www.empowermentproject.com/about/film/.
"Introduction to Learning Dynamical Systems." Computer Science Department, Brown University. http://cs.brown.edu/research/ai/dynamics/tutorial/Documents/DynamicalSystems.html.
Xu, April Xiaoyi. "Pomona Math Prof. Ami Radunskaya Elected President of the Association for Women in Mathematics" Pomona College. March 9, 2016. https://www.pomona.edu/news/2016/03/09-pomona-math-prof-ami-radunskaya-elected-president-association-women-mathematics.

INGRID DAUBECHIES

1. "Maths is (also) for women." The World Academy of Sciences, July 29, 2014, https://twas.org/article/maths-also-women.
2. Introduced in 1927, Heisenberg's uncertainty principle expresses the idea that the more precise a particle's position in space, the less precisely its momentum can be known. It is represented by the equation $\sigma_\chi \sigma_\rho \geq \frac{h}{2\pi}$, where σ_χ represents the standard deviation of a particle's position, σ_ρ represents the standard deviation of a particle's momentum, and $\frac{h}{2\pi}$ is the *reduced Planck constant* (also known as the *Dirac constant* after the English theoretical physicist Paul Dirac).
3. "Maths is (also) for women."
4. Ibid.
5. J.J. O'Connor and E.F. Robertson, "Ingrid Daubechies," School of Mathematics and Statistics, University of St Andrews, Scotland, http://www-groups.dcs.st-and.ac.uk/history/Biographies/Daubechies.html.

For more information on Ingrid Daubechies, consult the following:
Daubechies, Ingrid. "Using Mathematics to Repair a Masterpiece." *Quanta Magazine*, September 29, 2016. https://www.quantamagazine.org/using-mathematics-to-repair-a-masterpiece-20160929/.

"Difference between Fourier transform and Wavelets." Mathematics, Stack Exchange. https://math.stackexchange.com/questions/279980/difference-between-fourier-transform-and-wavelets.

"Ingrid Daubechies." Electrical and Computer Engineering, Duke University. http://ece.duke.edu/faculty/ingrid-daubechies.

Schlaefli, Samuel. "Using applied mathematics to track down counterfeits." ETH Zürich. October 21, 2015. https://www.ethz.ch/en/news-and-events/eth-news/news/2015/10/pauli-lectures.html

Sipics, Michelle. "The Van Gogh Project: Art Meets Mathematics in Ongoing International Study." SIAM News, Society for Industrial and Applied Mathematics. May 18, 2009. https://www.siam.org/news/news.php?id=1568.

Stancill, Jane. "Duke math professor wins $1.5 million award." The News & Observer, August 1, 2016. http://www.newsobserver.com/news/local/education/article93169462.html.

DAINA TAIMINA

1. Elizabeth Landau, "Geek Out!: Crochet sculptures teach higher math," *SciTechBlog*, CNN, April 30, 2010, http://scitech.blogs.cnn.com/2010/04/30/geek-out-crochet-sculptures-teach-higher-math/.

2. Michelle York, "Professor Lets Her Fingers Do the Talking," *The New York Times*, July 11, 2005, http://www.nytimes.com/2005/07/11/nyregion/professor-lets-her-fingers-do-the-talking.html.

3. Interview, December 2017.

4. "Daina Taimina," Department of Mathematics, Cornell University, http://www.math.cornell.edu/~dtaimina/base.html.

For more information on Daina Taimina, consult the following:

Artmann, Benno. "Euclidean geometry." *Encyclopedia Britannica*. https://www.britannica.com/topic/Euclidean-geometry.

"Daina Taimina." Blogger.com. https://www.blogger.com/profile/13079530989875080286.

Henderson, David W. and Daina Taimina. "Crocheting the Hyperbolic Plane." *Prepublication in Mathematical Intelligencer 23*, no. 2 (Spring 2001). http://www.math.cornell.edu/~dwh/papers/crochet/crochet.html.

Wertheim, Margaret, David Henderson, and Daina Taimina. "Crocheting the Hyperbolic Plane: An Interview with David Henderson and Daina Taimina." *Cabinet Magazine* no. 16 (Winter 2004/05). http://www.cabinetmagazine.org/issues/16/crocheting.php.

TATIANA TORO

Quotes from interview, December 2017.

For more information on Tatiana Toro, consult the following:

Arts & Sciences Web Team. "Tatiana Toro Named Chancellor's Professor at UC Berkeley." Department of Mathematics, University of Washington. December 10, 2016. https://math.

washington.edu/news/2016/12/10/tatiana-toro-named-chancellors-professor-uc-berkeley.

"Tatiana Toro." John Simon Guggenheim Memorial Foundation. https://www.gf.org/fellows/all-fellows/tatiana-toro/.

"Tatiana Toro." Lathisms. http://www.lathisms.org/wednesday-october-12th.html.

KAREN E. SMITH

1. Alexander Diaz-Lopez, "Karen E. Smith Interview," interview by Laure Flapan, *Notices of the AMS* 64 no. 7 (August 2017): 718–720, http://dx.doi.org/10.1090/noti1544.
2. Ibid.
3. Ibid.
4. In mathematics, singularities are the points at which a function is not defined, as in the case of $f(x) = 1/x$ at $x = 0$.
5. The term *cohomology* refers to a sequence of abelian groups (i.e., commutative algebraic structures) linked with a topological space (i.e., a set of points and associated neighborhoods, such as 3D manifolds).
6. For additional information on Smith's research interests, see http://www.math.lsa.umich.edu/~kesmith/research.pdf.
7. Presented in Seattle, Washington, in January 2016, the title of Smith's lecture was "The Power of Noether's Ring Theory in Understanding Singularities of Complex Algebraic Varieties."
8. Karen E. Smith, "An introduction to tight closure," September 26, 2012, https://arxiv.org/pdf/math/0209378.pdf.

For more information on Karen E. Smith, consult the following:

O'Connor, J.J. and E.F. Robertson. "Karen Ellen Smith." School of Mathematics and Statistics, University of St Andrews, Scotland. http://www-history.mcs.st-and.ac.uk/Biographies/Smith_Karen.html.

Riddle, Larry. "Karen E. Smith." Biographies of Women Mathematicians. Agnes Scott College. Last modified August 7, 2017. https://www.agnesscott.edu/lriddle/women/smithk.htm.

GIGLIOLA STAFFILANI

1. Billy Baker, "A life of unexpected twists takes her from farm to math department," *The Boston Globe*, April 28, 2008, http://archive.boston.com/news/science/articles/2008/04/28/a_life_of_unexpected_twists_takes_her_from_farm_to_math_department/.
2. Ibid.
3. Interview, October 2017.
4. Ibid.
5. Ibid.
6. Ibid.
7. Ibid.
8. Ibid.
9. Ibid.

10. Lavinia Pisani, "Gigliola Staffilani and her parable to success," *L'Italo Americano*, June 30, 2016, http://www.italoamericano.org/story/2016-6-30/staffilani.

11. Baker, A life of unexpected twists."

12. Interview, October 2017.

13. Ibid.

14. Alexander Diaz-Lopez, "Gigliola Staffilani Interview," *Notices of the AMS* 63, no. 11 (December 2016): 1250–1251, http://dx.doi.org/10.1090/noti1452.

15. Interview, October 2017.

For more information on Gigliola Staffilani, consult the following:

"Gigliola Staffilani: A Woman in Mathematics." Mathematical Association of America. May 16, 2008. https://www.maa.org/news/math-news/gigliola-staffilani-a-woman-in-mathematics.

"Gigliola Staffilani." Department of Mathematics, Massachusetts Institute of Technology. http://math.mit.edu/directory/profile.php?pid=262.

"Gigliola Staffilani." John Simon Guggenheim Memorial Foundation. https://www.gf.org/fellows/all-fellows/gigliola-staffilani/.

"Personal Profile of Prof. Gigliola Staffilani." Mathematical Sciences Research Institute. https://www.msri.org/people/3096.

ERICA WALKER

1. Erica N. Walker, *Building Mathematics Learning Communities* (New York: Teachers College Press, 2012): xiii–xiv.

2. Interview, October 2017.

3. Ibid.

4. "Dr. Erica N. Walker," Teachers College, Columbia University, http://www.tc.columbia.edu/faculty/walker/.

5. Eric A. Hurley, "Minority Postdoctoral Fellows," Teachers College Newsroom, Columbia University, May 20, 2002, http://www.tc.columbia.edu/articles/2001/november/minority-postdoctoral-fellows/.

6. Interview, October 2017.

7. "Dr. Erica N. Walker," Teachers College.

8. Interview, October 2017.

9. Ibid.

10. Evelyn Lamb, "Black Mathematical Excellence: A Q&A with Erica Walker," *Scientific American*, February 15, 2016, https://blogs.scientificamerican.com/roots-of-unity/black-mathematicians-erica-walker/.

For more information on Erica Walker, consult the following:

"Dr. Erica N. Walker: Professional Background." Teachers College, Columbia University. http://www.tc.columbia.edu/faculty/walker/profback.html.

Walker, Erica N. Building Mathematics Learning Communities. New York: Teachers College Press, 2012.

TRACHETTE JACKSON

1. Interview, October 2017.

2. Ibid.

3. Evelyn Lamb, "Mathematics, Live: A Conversation with Victoria Booth and Trachette Jackson," *Scientific American*, October 9, 2013, https://blogs.scientificamerican.com/roots-of-unity/mathematics-live-a-conversation-with-victoria-booth-and-trachette-jackson/.

4. Interview, October 2017.

5. Ibid.

6. Lamb, "Mathematics, Live."

7. Interview, October 2017.

For more information on Trachette Jackson, consult the following:

"Funded Grant: Trachette L. Jackson: Combining continuous and discrete approaches to study sustained angiogenesis associated with vascular tumor growth." James S. McDonnell Foundation. https://www.jsmf.org/grants/2005005/

"Trachette Jackson." Math Alliance. https://mathalliance.org/mentor/trachette-jackson/.

"Trachette Jackson." ScienceMakers, *The HistoryMakers: The Nation's Largest African American Oral History Collection*. October 22, 2012. http://www.thehistorymakers.org/biography/trachette-jackson.

Williams, Scott W. "Trachette Jackson." Black Women in Mathematics. Mathematics Department, State University of New York, Buffalo. http://www.math.buffalo.edu/mad/PEEPS/jackson_trachette.html.

CARLA COTWRIGHT-WILLIAMS

1. Interview, December 2017.

2. Ibid.

3. Ibid.

4. Ibid.

5. Interview, October 2017.

6. "Carla Cotwright," Mathematically Gifted and Black, February 9, 2017, http://www.mathematicallygiftedandblack.com/profiles/February_9.html.

7. Ibid.

8. Interview, October 2017.

9. Carla Cotwright-Williams, "Go BIG or go home? Go BIG." BIG Math Network, February 9, 2017, https://bigmathnetwork.wordpress.com/2017/02/09/blogpost-carla-cotwright-williams/.

10. Interview, October 2017.

11. Interview, December 2017.

12. Interview, October 2017.

13. Ibid.

14. Cotwright-Williams, "Go BIG or go home?"

15. Interview, October 2017.

16. Interview, December 2017.

For more information on Carla Cotwright-Williams, consult the following:

AMS Washington Office. "AMS Congressional Fellow Chosen." *Notices of the AMS* 59, no 7, August 2012. http://www.ams.org/notices/201207/rtx120700974p.pdf.

Davenport, Thomas H. and D.J. Patil. "Data Scientist: The Sexiest Job of the 21st Century." *Harvard Business Review*, October 2012. https://hbr.org/2012/10/data-scientist-the-sexiest-job-of-the-21st-century.

"Doctoral Alumna Uses Math for Public Good." Graduate School, The University of Mississippi. Summer 2017. https://gradschool.olemiss.edu/newsletter-summer-2017/doctoral-alumna-uses-math-for-public-good/.

EUGENIA CHENG

1. Nicola Davis, "Mathematician Eugenia Cheng: 'We hate having rules imposed on us,' " *The Guardian*, February 26, 2017, https://www.theguardian.com/science/2017/feb/26/eugenia-cheng-interview-observer-nicola-davis.

2. Interview, October 2017.

3. Davis, "Mathematician Eugenia Cheng."

4. Interview, October 2017.

5. Davis, "Mathematician Eugenia Cheng."

For more information on Eugenia Cheng, consult the following:

Angier, Natalie. "Eugenia Cheng Makes Math a Piece of Cake." *The New York Times*, May 2, 2016. https://www.nytimes.com/2016/05/03/science/eugenia-cheng-math-how-to-bake-pi.html.

Cheng, Eugenia. EugeniaCheng.com. http://eugeniacheng.com/.

Cheng, Eugenia. "Why I Don't Like Being a Female Role Model." *Bright Magazine*, July 7, 2015. https://brightthemag.com/why-i-don-t-like-being-a-female-role-model-10055873ea97.

MARYAM MIRZAKHANI

1. Erica Klarreich, "A Tenacious Explorer of Abstract Surfaces," *Quanta Magazine*, August 12, 2014, https://www.quantamagazine.org/maryam-mirzakhani-is-first-woman-fields-medalist-20140812/.

2. Ibid.

3. "Interview with Research Fellow Maryam Mirzakhani," 2008 Annual Report, Clay Mathematics Institute, http://www.claymath.org/library/annual_report/ar2008/08Interview.pdf.

4. Ibid.

5. Ibid.

6. Klarreich, "A Tenacious Explorer of Abstract Surfaces."

7. Andrew Myers And Bjorn Carey, "Maryam Mirzakhani, Stanford mathematician and Fields Medal winner, dies," *Stanford News*, July 15, 2017, https://news.stanford.edu/2017/07/15/maryam-mirzakhani-stanford-mathematician-and-fields-medal-winner-dies/.

8. If you're curious about string theory, Brian Greene's 2003 bestseller, *The Elegant Universe* (W.W. Norton & Company), is a great place to start.

9. "Interview with Research Fellow Maryam Mirzakhani."

10. Klarreich, "A Tenacious Explorer of Abstract Surfaces."

11. Ibid.

12. "The work of Maryam Mirzakhani," August 18, 2014, http://www.math.harvard.edu/~ctm/papers/home/text/papers/icm14/icm14.pdf.

13. Jordan Ellenberg, "Math Is Getting Dynamic," *Slate*, August 13, 2014, http://www.slate.com/articles/life/do_the_math/2014/08/maryam_mirzakhani_fields_medal_first_woman_to_win_math_s_biggest_prize_works.html.

14. Klarreich, "A Tenacious Explorer of Abstract Surfaces."

15. Myers and Carey, "Maryam Mirzakhani, Stanford mathematician."

16. "Iranian math genius Mirzakhani passes away," PressTV, July 15, 2017, http://www.presstv.com/Detail/2017/07/15/528535/Iran-Maryam-Mirzakhani-cancer-US.

For more information on Maryam Mirzakhani, consult the following:

Dehghan, Saeed Kamali. "Maryam Mirzakhani: Iranian newspapers break hijab taboo in tributes." *The Guardian*, July 16, 2017. https://www.theguardian.com/world/2017/jul/16/maryam-mirzakhani-iranian-newspapers-break-hijab-taboo-in-tributes/

O'Connor, J.J. and E.F. Robertson. "Maryam Mirzakhani." School of Mathematics and Statistics, University of St Andrews, Scotland. http://www-history.mcs.st-and.ac.uk/Biographies/Mirzakhani.html.

Roberts, Siobhan. "Maryam Mirzakhani's Pioneering Mathematical Legacy." *The New Yorker*, July 17, 2017. https://www.newyorker.com/tech/elements/maryam-mirzakhanis-pioneering-mathematical-legacy.

CHELSEA WALTON

1. "Donald in Mathmagic Land," Wikiquote, https://en.wikiquote.org/wiki/Donald_in_Mathmagic_Land.

2. "Chelsea Walton," Mathematically Gifted and Black, February 25, 2017, http://www.mathematicallygiftedandblack.com/profiles/February_25.html.

3. Ibid.

4. Ibid.

5. Eryn Jelesiewicz, "Temple mathematician Chelsea Walton named a 2017 Sloan Research Fellow," Temple Now, Temple University, March 7, 2017, https://news.temple.edu/news/2017-03-07/chelsea-walton-2017-sloan-research-fellow.

6. Rae Paoletta, "These Black Female Mathematicians Should Be Stars in the Blockbusters of Tomorrow," *Gizmodo*, March 8, 2017, https://gizmodo.com/these-black-female-mathematicians-should-be-stars-in-th-1792636094.

7. Alexander Diaz-Lopez, "Chelsea Walton Interview," *Notices of the AMS* 65 no. 2 (February 2018): 164–166. http://dx.doi.org/10.1090/noti1631.

For more information on Chelsea Walton, consult the following:

"Chelsea Walton: Bio." Mathematics, Temple University. https://math.temple.edu/~notlaw/bio.html.

"Chelsea Walton: CV." Department of Mathematics, Massachusetts Institute of Technology. Last modified January 6, 2015. http://math.mit.edu/~notlaw/CV3.pdf.

PAMELA E. HARRIS

Quotes from interview, December 2017.

ACKNOWLEDGMENTS

I would like to thank all my mentors who helped me along in my journey. Thank you also to Jeannine Dillon and Melanie Madden at Quarto Publishing, who invited me to share my story and those of the women in this book. Also, I could not have done this book without the careful editing of Stephanie Graham, Erin Canning, Jason Chappell, and Madeleine Vasaly. Thank you to Merideth Harte, Jen Cogliantry, and Phil Buchanan for designing and laying out this beautiful book quickly. Thanks also to illustrator Sean Yates and photo researcher Lesley Hodgson. Without all of you, this book would not have been possible!

Finally, I dedicate this book to Mom and my husband, Donald: without your unconditional love and support, I would not be where I am today.

IMAGE CREDITS

ALAMY

© dpa picture alliance archive: 75, 126 (top), 131; © Everett Collection Inc: 98 (bottom); © Gado Images: 68 (top); © Heritage Image Partnership Ltd: 30; © Pictorial Press Ltd: 13 (center left); © Science History Images: 20 (top), 26 (top), 40

COURTESY SYLVIA BOZEMAN

100, 102, 103 (top and bottom), and 105-107

CATHOLIC UNIVERSITY ARCHIVES

© (Haynes-Lofton Family Collection Box 67) The American Catholic History Research Center and University Archives (ACUA): 44

COURTESY EUGENIA CHENG

172 (bottom); © Paul Crisanti of PhotoGetGo: 176; © RoundTurnerPhotography.com: 172 (top)

COLUMBIA UNIVERSITY

© Gift of the Wellesley College Class of 1883, Zeta Chapter of Phi Delta Gamma, and the Women's Graduate Club of Columbia University. Art Properties, Avery Architectural & Fine Arts Library: 34; © Historical Photo Collection, Box 84, University Archives, Rare Book and Manuscript Library: 32 (top)

COURTESY CARLA COTWRIGHT-WILLIAMS

166, 168 (top); © Bryan Williams: 171

GETTY IMAGES

© ATTA KENARE/AFP: 185; © Bettmann: 43; © Hulton Archive: 61 (bottom); © James Whitmore/The LIFE Images Collection: 62; © Jeffrey R. Staab/CBS: 175; © Leigh Vogel/WireImage: 95 (top); © Library of Congress/Interim Archives: 54 (top); © Maryam Mirzakhani/NASA 382199/Corbis: 178 (top); © Samsung / Barcroft Images / Barcroft Media: 174; © SSPL: 17, 73

COURTESY PAMELA HARRIS

192, 195; "Combinatorial problems related to Kostant's weight multiplicity formula," Ph.D. thesis, University of Wisconsin-Milwaukee: 193, 194

COURTESY HARVEY MUDD COLLEGE/MARIA KLAWE

114, 116–119

COURTESY FERN HUNT

108, 113

THE IMAGE WORKS

© DPA/SOA

SMITHSONIAN INSTITUTION

© Grace Murray Hopper Collection, Archives Center, National Museum of American History: 56, 57 (top)

COURTESY DAINA TAIMINA

101, 134, 137

COURTESY TATIANA TORO

138, 141

U.S. CENSUS BUREAU

57 (bottom)

U.S. DEPARTMENT OF DEFENSE

U.S. Air Force: 63 (top); U.S. Navy: 50 (bottom); 58, 59, 63 (bottom), 64

© GINI WADE

16 (TOP) COURTESY / © WGBH EDUCATIONAL FOUNDATION:

9

COURTESY WIKIMEDIA FOUNDATION

12, 20 (bottom), 21, 26 (bottom), 28, 36–38, 42, 52 (top), 54 (bottom), 110, 123, 125, 126 (bottom), 128, 130, 145, 152, 157, 169, 177, 180, 182, 187; Archive of the Berlin-Brandenburg Academy of Sciences and Humanities: 21 (top), 29 (top); Charles Marville/National Gallery of Art, Washington, Patrons' Permanent Fund: 25; Daphne Weld Nichols: 92 (top); DLR German Aerospace Center: 81; Douglas W. Reynolds: 77; Draper Laboratory, restored by Adam Cuerden: 94; Harvard College Observatory: 33, 50 (top); Internet Archive: 15, 23 (bottom), 24 (top), 186 (bottom); Johannes Meiner, ETH Library: 41; Killivalavan Solai: 191; Manfred Kuzel: 47; Nationalmuseum, photographed by Erik Cornelius: 29 (bottom); RMN-Grand Palais (Château de Versailles)/Gérard Blot: 24 (bottom); State Tretyakov Gallery: 27 (top); Stefan Zachow of the International Mathematical Union, retouched by King of Hearts: 178 (bottom); Tiia Monto: 149; Trocaire: 98 (top); TURNBULL WWW SERVER, School of Mathematical and Computational Sciences, University of St Andrews: 27 (bottom); Voice of America/Ali Shaker: 99; William T. Ziglar, Jr: 67

COURTESY TALITHIA WILLIAMS

6; © Tableau: 7

Other Images: © Ryan Brandenberg/Temple University: 186 (top); © Bruce Gilbert: 156; © Byron Hooks/Lat34north: 104 (bottom); © Richard J Misch: 87; Courtesy Pomona College/Ami Radunskaya: 120 (top); Schomburg Center for Research in Black Culture/New York Public Library: 66 (bottom); Courtesy Karen Smith: 144; Courtesy Gigliola Staffilani: 150; Courtesy Vaughan Family: 66 (top)

INDEX